TC 3-21.90

I0150660

Mortar Platoon
Collective Task Publication

August 2013

Headquarters, Department of the Army

Training Circular
No. 3-21.90

Headquarters
Department of the Army
Washington, DC, 2 August 2013

Mortar Platoon
Collective Task Publication

Contents

Figures

Tables

Preface

Purpose

The training circular (TC) is a tool that platoon leaders can use as an aid during training strategy development. The products in this TC are developed to support the battalion's mission plan.

Scope

This TC provides guidance for platoon leaders, leaders, and Soldiers who are responsible for planning, preparing, executing, and assessing training of the mortar platoon.

Applicability

This publication applies to the Active Army, the Army National Guard (ARNG), Army National Guard of the United States (ARNGUS), and the United States Army Reserve (USAR) unless otherwise stated.

Intended Audience

The primary target audience for this TC is the platoon leader and other leaders within a mortar platoon. The secondary audience comprises training developers involved in developing training support materials for professional military education.

Feedback

The proponent for this publication is the U.S. Army Training and Doctrine Command. The preparing agency is the U.S. Army Maneuver Center of Excellence. Send comments and recommendations by any means, U.S. mail, e-mail, fax, or telephone, following DA Form 2028, *Recommended Changes to Publications and Blank Forms*. More information is available by phone. Point of contact information is as follows.

E-mail:	BENN.CATD.DOCTRINE@CONUS.ARMY.MIL
Phone:	COM 706-545-7114 or DSN 835-7114
Fax:	COM 706-545-8511 or DSN 835-8511
U.S. Mail:	Commanding General, Maneuver Center of Excellence
	Directorate of Training and Doctrine
	Doctrine and Collective Training Division
	ATTN: ATZB-TDD
	Fort Benning, GA 31905-5410

Unless otherwise stated in this publication, masculine nouns and pronouns refer to both men and women.

This page intentionally left blank.

Chapter 1
Introduction

Mortars are suppressive indirect fire weapons. They can be employed to neutralize or destroy area or point targets, screen large areas with smoke, and provide illumination or coordinated high-explosive illumination. The mortar platoon's mission is to provide close and immediate indirect fire support for maneuver battalions and companies. (Refer to FM 3-22.90.)

SECTION I – TEXT REFERENCES

1-1. Table 1-1 contains the references used in this chapter.

Table 1-1. Guide for subjects referenced in text

Reference	Subject
ATS	Deputy Chief of Staff, G-3/5/7 memorandum, *Army Training Strategy*
LDS	*The Army Leader Development Strategy for a 21st Century Army*
FM 3-22.90	*Mortars*
ATLDG	Army, G-3/5/7 memorandum, *Army Training and Leader Development Guidance*, FY 10-11
ADP 3-0	*Unified Land Operations*
FM 3-21.20	*Infantry Rifle Battalion*
FM 3-90.5	*The Combined Arms Battalion*
ADP 7-0	*Training Units and Developing Leaders*
FM 6-22	*Army Leadership: Competent, Confident, and Agile*
ATN	*Army Training Network* link: https://atn.army.mil/index.aspx
FM 1-02	*Operational Terms and Graphics*
ADP 6-0	*Mission Command*

Table 1-1. Guide for subjects referenced in text (continued)

Reference	Subject
AR 350-1	*Army Training and Leader Development*

SECTION II – ARMY APPROACH TO TRAINING

1-2.   Before platoon leaders plan, prepare, execute, and assess training, they first must have a clear understanding of the Army's training and leader development strategies, training system, and training management.

ARMY TRAINING STRATEGY

1-3.   The Army goal is to routinely generate trained and ready units for both current missions and future contingencies at an operational tempo that is sustainable (*Army Training and Leader Development Guidance* (ATLDG), FY 10-11). To accomplish this goal, the Army G-3/5/7 has developed the comprehensive Army training strategy (ATS).

1-4.   The ATS describes the ends, ways, and means required to adapt Army training programs to an era of persistent conflict, to prepare units and leaders to conduct decisive action operations, and to rebuild strategic depth. The ATS generates cohesive, trained, and ready forces that can dominate at any point in the spectrum of conflict, in any environment, and under all conditions.

1-5.   The ATS has identified 10 goals. Each goal has supporting objectives that details the ATS. Obtaining each goal ensures the Army generates trained and ready units. The goals are:

- Train units for decisive action operations.
- Enable adaption of training.
- Train and sustain Soldier skills.
- Train and sustain Army civilian skills.
- Sustain and improve effectiveness of combat training centers (CTCs).
- Provide training at home station and while deployed.
- Provide training support system live, virtual, constructive, and gaming (LVCG) enablers.
- Increase culture and foreign language competencies.
- Provide supporting and integrating capabilities.
- Resource the Army training strategy.

ARMY LEADERSHIP DEVELOPMENT STRATEGY

1-6.   While the ATS was being developed, the commanding general (CG) of the U.S. Training and Doctrine Command (TRADOC) concurrently developed a leader development strategy (LDS). *The Army Leader Development Strategy for a 21st Century Army* discusses how the Army will adapt the way in which it develops leaders. This strategy presents the challenges of the operational environment (OE), the implications of the OE on leader development, and the mission, framework, characteristics, and imperatives of, and how to implement the strategy. The LDS describes eight specific imperatives designed to guide the policy and actions necessary to produce the future leaders the Army will need.

1-7.   The LDS is part of a campaign of learning. It seeks to be as adaptive and innovative as the leaders it must develop. The LDS is grounded in Army leadership doctrine and seeks to deliver the leader qualities described in both Army doctrine and capstone concepts. (Refer to FM 6-22 and *The Army Leader Development Strategy for a 21st Century Army*.) The following documents describe leadership qualities:

- **ADP 3-0**. This manual discusses decisive action, which includes the elements of offensive, defensive, and stability (or defense support of civil authorities. Army forces conduct decisive and sustainable land operations through the simultaneous combination of offensive, defensive, and stability operations (or defense support of civil authorities) appropriate to the mission and environment. Army forces conduct regular and irregular warfare against both conventional and hybrid threats.

ARMY TRAINING SYSTEM

1-8.   The Army Training System prepares Soldiers, organizations, and their leaders to conduct decisive action operations. The training system is built upon a foundation of disciplined, educated, and professional Soldiers and leaders, adhering to principles that provide guidance.

Principles of Unit Training

1-9.   Army training exists today in many domains and throughout the OE in which Soldiers and leaders find themselves. To maintain a professional baseline the Army has developed 11 training principles that govern Army training. The principles provide a broad but basic foundation to guide how commanders and other leaders plan, prepare, execute, and assess effective training. (Refer to ADP 7-0.) The 11 principles of training are:

- **Commanders and other leaders are responsible for training**. Commanders are ultimately responsible for the training, performance, and readiness of their Soldiers. However, leaders across all echelons and throughout the operational Army and generating force are responsible for training their respective organizations.
- **Noncommissioned officers train individuals, crews, and small teams**. Noncommissioned officers (NCOs) are the primary trainers of enlisted Soldiers, crews, and small teams. Officers and NCOs have a special training relationship; their training responsibilities complement each other. This relationship spans all echelons and types of organizations. Noncommissioned officers are usually an organization's most experienced trainers.
- **Train to standard**. Army training is performed to standard. Leaders prescribe tasks with their associated standards that ensure their organization is capable of accomplishing its doctrinal or assigned mission. A standard is the minimum proficiency required to accomplish a task under a set of conditions.
- **Train as you will fight**. "Fight" includes lethal and nonlethal skills in decisive action operations. "Train as you will fight" means training under the conditions of expected, anticipated, or plausible OEs.
- **Train while operating.** Training continues when a unit is engaged in operations. Combat builds experience, but not necessarily effectiveness. To adapt to constantly changing situations, units continue to train even in the midst of campaigns. Platoon leaders use available time to rehearse mission execution and prepare for likely contingencies. They conduct after-action reviews (AARs) after completing operations—and after completing intermediate tasks—to capture lessons learned for future operations.
- **Train fundamentals first**. Platoons train their most important collective tasks first; the tasks that are basic to mission proficiency.
- **Train to develop operational adaptability.** Although planning is critical to successful training, circumstances may cause plans to change. Leaders prepare for personnel turbulence and equipment shortages even though the Army Force Generation (ARFORGEN) system tries to ensure personnel and equipment objectives are met before training

begins. Platoon leaders develop training, manning, and equipping contingency plans. They train their Soldiers to assume other positions on short notice.

- **Understand the operational environment.** Commanders understand the OE and how it affects training. They replicate operational conditions, including anticipated variability, in training. Platoon leaders must understand and use the operational variables: political, military, economic, social, information, infrastructure, physical environment, and time (PMESII–PT) and the mission variables: mission, enemy, terrain and weather, troops and support available, time available, and civil considerations (METT–TC) to assist the commander in analyzing the OE and mission as they plan, prepare, execute, and assess training.

- **Train to sustain.** Platoons must be capable of operating continuously while deployed. Essential for continuous operations, sustainment is an integral part of training.

- **Train to maintain.** Platoon leaders allocate time for Soldiers to maintain themselves and their equipment to standard during training events. This time includes scheduled and routine equipment maintenance periods and assembly area operations. Leaders train their subordinates to appreciate the importance of maintaining their equipment. Organizations tend to perform maintenance during operations to the standards they practice in training.

- **Conduct multiechelon and concurrent training.** Multiechelon training is a technique that allows for the simultaneous training of more than one echelon on different or complementary tasks. It is the most efficient way to train, especially with limited resources. It requires synchronized planning and coordination by commanders and other leaders at each affected echelon.

Principles of Leader Development

1-10. Leader development is deliberate, continuous, and progressive, spanning a leader's entire career. Leader development comprises training and education gained in schools; the learning and experiences gained while assigned to organizations; and the individual's own self-development.

1-11. Platoon leaders are responsible for the professional development of subordinate leaders, and for building and sustaining the leader characteristics and skills. (Refer to FM 6-22.) Platoon leaders are responsible for leader development of subordinates and are every leader's top priority. Effective training and education build good leaders, and good leaders develop and execute effective training and education in schools and units. The experience gained during assignments puts the training and education into practice and provides the skills and knowledge leaders need to be versatile, adaptable, well-rounded, competent professionals. The Army's principles of leader development are:

- **Lead by example**. Platoon leaders are role models. To demonstrate good leadership is to teach good leadership. Everything a leader does and says is scrutinized, analyzed, and often imitated. The example set by platoon leaders influences the thoughts and attitudes of their subordinates, their families, and their peers. A good example positively influences the development of subordinates.

- **Take responsibility for developing subordinate leaders.** Platoon leaders take responsibility for developing their subordinate leaders. They directly observe, assess and provide honest informal and formal feedback to their subordinates. They discuss ways to sustain and improve leader skills, knowledge, abilities, and behaviors with their subordinate leaders as often as needed. They ensure subordinates undergo experiences that enhance their skills, knowledge, abilities, and behaviors; prepare them for success; improve their adaptability; and prepare them for future responsibilities. They ensure their subordinates attend professional military education at the right time in their careers and functional training to make them effective leaders in their units of assignment

- **Create a learning environment for subordinate leaders**. Leaders learn in an environment conducive to growth. Growth occurs best in environments that provide subordinates with opportunities to overcome obstacles and make difficult decisions. Platoon leaders encourage their subordinates to seek challenging assignments, and platoon leaders underwrite subordinates' honest mistakes. Learning comes from both successes and failures. Leaders must feel comfortable taking risks and trying new approaches to training. An environment that allows subordinate leaders to make honest—as opposed to repeated or careless—mistakes without prejudice is essential to leader development.

- **Train leaders in the art and science of mission command**. Platoon leaders approach mission command training from two perspectives. First, they train themselves and their subordinates on how to conduct operations using mission command. (Refer to ADP 3-0 and ADP 6-0.) Second, they follow the principles of mission command in training management. Specifically, they tell their subordinates the purpose for training and the end state they expect from it, but they leave the determination of how to achieve the end state to the subordinate. As appropriate, they provide guidance requested by the subordinate leader. Employing mission command in training follows the principle of train as you will fight. Using mission command principles improves not only mission command skills, but it also encourages risk-taking, initiative, and creativity.

- **Train to develop adaptive leaders.** The Army continues to succeed under the most challenging conditions because Soldiers adapt to unexpected situations. Operational adaptability begins in the schools and is then put into practice during tough, realistic training situations, well before leaders are engaged in decisive action. Knowing that change will occur, effective platoon leaders plan for it and develop potential contingency plans to mitigate the effects of change. Effective platoon leaders also look for indicators that change is about to occur so they can ease the transition effects. Placing subordinate leaders into changing, unfamiliar, and uncomfortable situations in training helps foster operational adaptability. The lessons they learn help develop intuition, confidence, and the ability to think on their feet. The Army trains leaders for their next position before they assume it. Cross-training provides unit depth and flexibility and builds leader confidence.

- **Train leaders to think critically and creatively.** The Army develops leaders able to solve difficult, complex problems. Leaders should be able to recognize the issue, quickly ask the right questions, consider a variety of alternative solutions, and develop effective solutions. They should be comfortable making decisions with minimal information. (Refer to ADP 5-0.)

- **Train leaders to know their subordinates and their families.** Every platoon leader should know his subordinates strengths, weakness, and capabilities. An effective leader maximizes a subordinate's strengths and helps him overcome weaknesses. Similarly, an effective leader provides advice, counsel, and support as subordinate leaders develop their own subordinates. Family well-being is essential to unit and individual readiness. The Army trains leaders to know and help not only the subordinates, but also their families. Training ensures subordinate leaders recognize the importance of families and are adept at helping individuals solve family issues and sustain sound relationships.

UNIT TRAINING MANAGEMENT

1-12. Unit training management (UTM) is the process used by Army leaders to identify training requirements and subsequently plan, prepare, execute, and assess training. UTM provides a systematic way of managing time and resources and of meeting training objectives through purposeful training activities.

1-13. The platoon leader's role in training is using the mission as the foundation. The platoon leader assists the troop commander in determining the tasks that the platoon will train. Platoon leaders must understand the unit's mission and the expected operational conditions to replicate in training. The commander identifies collective tasks to train and the associated risks of not training other collective tasks to proficiency, and relays that knowledge to the platoon leaders.

1-14. The conditions are either the ones described in the higher unit's training and leader development guidance, or those likely to be encountered in a mission. The platoon leader visualizes the platoon's required state of readiness for the mission and the training necessary to achieve mission proficiency, given the platoon leader's assessment of current task proficiency. The platoon leader describes the training plan in training and leader development guidance or operation orders and directs its execution to subordinate leaders. By participating in, and overseeing training and listening to feedback from subordinates, unit platoon leaders assess the platoon's task proficiency and whether the training being conducted contributes to mission readiness.

1-15. UTM is the practical application of the training doctrine found in ADP 7-0. The UTM information contained in ADP 7-0 supersedes FM 7-1. UTM provides a systematic way of managing time and resources and of meeting training objectives through purposeful training activities.

1-16. ADP 7-0 and UTM are posted within the Army Training Network (ATN). The ATN is an Internet Web site that provides best practices, examples, tools and lessons learned.

1-17. These references are linked and designed to be used in concert as a digital resource. ADP 7-0 provides the intellectual framework of what Army training is, while UTM provides the practical how-to of planning, preparing, executing, and assessing training in detail. The ATN, as the digital portal to both documents, additionally provides a wealth of other training resources to include the latest training news, information, products and links to other Army training resources.

ARMY FORCE GENERATION

1-18. Army force generation (ARFORGEN) is a process that progressively builds unit readiness over time during predictable periods of availability to provide trained, ready, and cohesive units prepared for operational deployments. (Refer to ADP 7-0).

1-19. Army force generation drives UTM within the Army. (Refer to ADP 7-0 for more information.) Training management is the process used by Army leaders to identify training requirements and subsequently plan, prepare, execute, and assess training.

1-20. The Army prepares and provides campaign capable, expeditionary forces through ARFORGEN, which applies to Regular Army (RA) and Reserve Component (RC) units (Army National Guard and U.S. Army Reserve).

1-21. Army force generation takes each unit through a three-phased readiness cycle (known as pools): reset, train/ready, and available. The reset, train/ready, and available force pools provide the framework for the structured progression of increased readiness in ARFORGEN. The force pools are defined as follows:

- **Reset force pool.** Units enter the reset force pool when they redeploy from long-term operations or complete their window for availability in the available force pool. The RA units remain in the reset force pool for at least six months and RC units remain in the reset force pool for at least 12 months. Units in the reset force pool have no readiness expectations.
- **Train/ready force pool.** A unit enters the train/ready force pool following the reset force pool. The train/ready force pool is not of fixed duration. Units in the train/ready force pool will increase training readiness and capabilities as quickly as possible, given the resource availability. Units may receive a mission to deploy during the train/ready force pool.

- **Available force pool.** Units focus on deployment and training to sustain METL fundamentals and correct any operational deficiencies. Units in the available force pool are at the highest state of training and readiness capability and are ready to deploy when directed. The available force pool window for availability is one year.

1-22. Units move from the available force pool to the reset force pool following a deployment or the end of their designated window of availability.

SECTION III – OTHER TRAINING CONSIDERATIONS

1-23. When implementing the platoon training plan, the platoon leader can use this TC for several specific purposes:

- Apply tasks, conditions, and standards of performance to the platoon's training objectives.
- Evaluate the platoon's ability to perform specific tasks.
- Examine the effectiveness of the training in preparing the platoon for future training and operations.
- Provide input for planning and resourcing training activities at the platoon level.

1-24. The specific details of each platoon's training plan varies depending on a variety of factors, including the following:

- The higher unit's METL.
- Training directives and guidance established by the chain of command.
- The platoon's training priorities.
- Available training resources, including training areas.

OPERATIONAL ENVIRONMENT

1-25. The operational environment is a composite of the conditions, circumstances, and influences that affect the employment of capabilities and bear on the decisions of the commander (JP 1-02). Army leaders plan, prepare, execute, and assess operations by analyzing the operational environment in terms of the operational variables and mission variables. The operational variables consist of political, military, economic, social, information, infrastructure, physical environment, time (known as PMESII-PT). The mission variables consist of mission, enemy, terrain and weather, troops and support available, time available, civil considerations (known as METT-TC). How these variables interact in a specific situation, domain (land, maritime, air, space, or cyberspace), area of operations, or area of interest describes a commander's operational environment but does not limit it. No two operational environments are identical, even within the same theater of operations, and every operational environment changes over time. Because of this, Army leaders consider how evolving relevant operational or mission variables affect force employment concepts and tactical actions that contribute to the strategic purpose.

OPERATIONAL VARIABLES

1-26. Battalion commanders and mortar platoon leaders analyze and describe the OE in terms of operational variables. Operational variables are those broad aspects of the environment, both military and nonmilitary, that may differ from one operational area to another and affect campaigns and major operations.

1-27. These operational variables are easily remembered using PMESII-PT.

MISSION VARIABLES

1-28. Army forces use mission variables to focus analysis on specific elements of the environment that apply to their mission. Upon receipt of a warning order or mission, Army tactical leaders narrow their focus to six mission variables. Mission variables are those aspects of the operational environment that directly affect a mission. They outline the situation as it applies to a specific Army unit. Mission variables are mission, enemy, terrain and weather, troops and support available, time available, and civil considerations (METT-TC). (Refer to ADP 3-0.) The platoon leader carefully analyzes the higher unit's OPORD to identify the platoon's purpose; the specified, implied, and essential tasks it must perform, and the time line for accomplishing those tasks. The following outline of METT-TC factors assists the platoon leader in analyzing the mission and creating a time line.

1-29. METT-TC is a memory aid that identifies these mission variables:

- **Mission.** The mission is the task, together with the purpose, that clearly indicates the action to be taken and the reason therefore. (Refer to JP 1-02.) Leaders analyze a mission in terms of specified tasks, implied tasks, and the commander's intent two echelons up.

- **Enemy.** This analysis includes not only the known enemy but also other threats to mission success. These include threats posed by multiple adversaries with a wide array of political, economic, religious, and personal motivations.

● **Terrain and weather.** Terrain and weather are natural conditions that profoundly influence operations. Terrain and weather are neutral; they favor neither side unless one is more familiar with—or better prepared to operate in—the physical environment. The platoon leader analyzes the terrain using the factors of terrain: observation, avenues of approach, key and decisive terrain, obstacles, and cover and concealment (OAKOC). Elements of the OAKOC and weather analysis include the following:

- Observation and fields of fire.
- Avenues of approach.
- Key terrain.
- Obstacles.
- Cover and concealment.
- Weather. Climate and weather can significantly impact military operations. For military applications, the term "weather" implies weather forecast information designed to support a planned future operation. (Refer to FM 2-01.3.) The following are military aspects of weather:
  - Visibility.
  - Wind.
  - Precipitation.
  - Cloud cover/ceiling.
  - Temperature.
  - Humidity.
  - Atmospheric pressure (as required).

● **Troops and support available.** Troops and support available are the number, type, capabilities, and condition of available friendly troops and support. These include resources from joint, interagency, multinational, host-nation, commercial (via contracting), and private organizations. They also include support provided by civilians.

● **Time available.** Time is critical to all operations. Controlling and exploiting it is central to initiative, tempo, and momentum. By exploiting time, leaders can exert constant pressure, control the relative speed of decisions and actions, and exhaust enemy forces. As part of this analysis, the platoon leader conducts reverse planning to ensure that all essential, specified, and implied tasks can be accomplished in the time available. He develops a reverse planning schedule (time line) beginning with actions on the objective and working backward through each step of the operation and preparation to the present time. This process also helps the platoon in making efficient use of planning and preparation time.

- **Civilian considerations.** Civil considerations are the influence of manmade infrastructure, civilian institutions, and attitudes and activities of the civilian leaders, populations, and organizations within an AO on the conduct of military operations. (Refer to ADP 6-0.) Most of the time, units are surrounded by noncombatants. These noncombatants include residents of the area of operation (AO), local officials, and governmental and NGOs. Based on information from higher HQ and their own knowledge and judgment, platoon leaders identify civil considerations that affect their mission.

THREATS

1-30. Threats facing U.S. forces today vary. They are not always enemy forces dressed in uniforms who are easily identified as foe, aligned on a battlefield, and opposite U.S. forces. Threats are nation-states, organizations, people, groups, or conditions that can damage or destroy life, vital resources, or institutions.

1-31. Threats are described in four major categories or challenges: traditional, irregular, catastrophic, and disruptive. While helpful in describing the threats the Army is likely to face, these categories do not define the nature of the adversary. Adversaries may use any and all of these challenges in combination to achieve the desired effect against the U.S. (Refer to ADP 3-0). The four threats are defined as follows:

- **Traditional.** States employing recognized military capabilities and forces in understood forms of military competition and conflict.
- **Irregular.** Opponent employing unconventional, asymmetric methods and means to counter traditional U.S. advantages.
- **Catastrophic.** Enemy that involves the acquisition, possession, and use of weapons of mass destruction and effects.
- **Disruptive.** Enemy using new technologies that reduce U.S. advantages in key operational domains.

Hybrid Threats

1-32. The term "hybrid threat" has recently been used to capture the seemingly increased complexity of operations and the multiplicity of actors involved. While the existence of innovative enemies is not new, today's hybrid threats demand that the platoon prepare for a range of possible threats simultaneously.

1-33. Hybrid threats are characterized by the combination of regular forces governed by international law, military tradition, and custom with irregular forces that are unregulated and as a result act with no restrictions on violence or targets for violence. This could include militias, terrorists, guerillas, and criminals. Such forces combine their abilities to use and transition between regular and irregular tactics and weapons. These tactics and weapons enable hybrid threats to capitalize on perceived vulnerabilities making them particularly effective.

DECISIVE ACTION

1-34. A mortar platoon operates in a framework of decisive action. ADP 3-0 provides a discussion of decisive action which includes the elements of offensive, defensive, and stability operations (or defense support of civil authorities).

1-35. Army forces conduct decisive and sustainable land operations through the simultaneous combination of offensive, defensive, and stability operations (or defense support of civil authorities) appropriate to the mission and environment. Army forces conduct regular and irregular warfare against both conventional and hybrid threats. The primary operations are:

- **Offensive.** Offensive operations are operations conducted to defeat and destroy enemy forces and seize terrain, resources, and population centers. They include movement to contact, attack, exploitation, and pursuit.
- **Defensive. .** Defensive operations are operations conducted to defeat an enemy attack, gain time, economize forces, and develop conditions favorable for offensive and stability tasks. These operations include mobile defense, area defense, and retrograde.
- **Stability operations (or defense support of civil authorities).** Stability operations are military missions, tasks, and activities conducted outside the United States to maintain or reestablish a safe and secure environment and to provide essential governmental services, emergency infrastructure reconstruction, and humanitarian relief. They include five tasks: establish civil security, establish civil control, restore essential services, support to governance, and support to economic and infrastructure development. Homeland defense and defense support of civil authorities represent Department of Defense support to U.S. civil authorities for domestic emergencies, law enforcement support, and other domestic activities, or from qualifying

entities for special events. They include the tasks: provide support for domestic disasters; provide support for domestic chemical, biological, radiological, nuclear, and high-yield explosives incidents; provide support for domestic civilian law enforcement agencies; and provide other designated support.

1-36. The simultaneous conduct of decisive action requires careful assessment, prior planning, and unit preparation as commanders shift their combinations of decisive action. For further information on decisive action refer to ADP 3-0.

MISSION-ESSENTIAL TASK LIST

1-37. A mission-essential task list (METL) is a list of collective tasks a unit must be able to perform successfully to accomplish its doctrinal or directed mission. (Refer to ADP 7-0.) To meet the demands of decisive action, the Headquarters, Department of the Army (HQDA) has standardized METLs for brigades and above. This standardization ensures that like units deliver the same capabilities and gives the Army the strategic flexibility to provide trained and ready forces to operational-level commanders.

TASK LIST DEVELOPMENT

1-38. The platoon leader starts with reviewing the higher unit's METL and training guidance. The platoon leader and the platoon sergeant with guidance from the battalion commander determine what collective tasks, individual tasks, battle/crew drills, and leader tasks are that support the battalion METL. (See Figure 1-1.) The platoon leader should include squad and section leaders in this task selection process as well. Based on the platoon leader's analysis and identification of tasks that supports the battalion METL, the platoon leader determines a training focus that supports the unit commanders training guidance. At the completion of METL review, the platoon leader has determined:

- Collective tasks that support the battalion METL.
- Individual tasks that support the collective tasks.
- Resources required for training to standards.

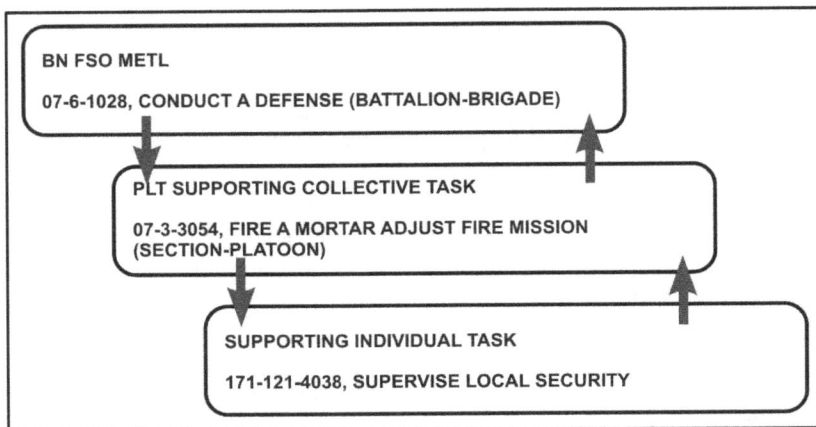

Figure 1-2. Collective and Individual tasks supporting higher unit METL

Platoon Leader Analysis

1-39. The platoon leader initiates the collective task, battle drill, crew drill, and leader task identification process with an analysis of the battalion collective tasks, any battle drills, battalion METL implementation guidance and training guidance. The platoon leader then identifies:

- Collective tasks, battle drills, crew drills, individual, and leader tasks the platoon will train.
- Collective tasks the platoon will not train and the risk for not training.
- An estimate of the time required to train.
- The conditions to train.
- Resources required.

Identify Collective Tasks

1-40. The platoon leader identifies the collective tasks, battle drills, crew drills, and leader tasks to train and the estimated time required to train to proficiency. Additionally, the platoon leader identifies those tasks the platoon can accept risk for not training.

Identify the Conditions

1-41. The platoon leader gains an understanding of the operational environment that the platoon will operate in and try to replicate the training conditions if possible. The conditions determine what resources are needed to re-create the OE.

1-42. The platoon leader and platoon sergeant determines the scarce and unique resources needed to train the selected collective tasks and individual tasks in the conditions previously identified. The platoon leader identifies those resources that require assistance from the battalion commander to obtain. Identifying these requirements first gives the battalion commander time for arranging and de-conflicting resources or finding alternatives.

Platoon Leader's Dialog

1-43. The battalion commander approves the platoon supporting collective task list. The approval normally occurs during the platoon leader's dialog. The dialog is a professional discussion between the platoon leader, platoon sergeant, company commander, and first sergeant (1SG) that sets the expectations for developing a platoon training plan. The platoon leader's dialog is the culminating point of task selection. In general this event:

- Is conducted face-to-face.
- Sets expectations for planning platoon training.
- Identifies any training readiness problems or risks.
- Sets expectations for the development of the platoon training plan.
- Identifies the training risks for those tasks not trained.

1-44. Upon completion of this dialog, the platoon leader has the necessary products to publish the platoon collective task, battle drill, and leader tasks.

Implementation Guidance

1-45. The platoon leader and platoon sergeant issues a document to the NCOs who summarize the platoon leader's dialog with the battalion commander. This is done primarily face-to-face. It provides the NCO's and Soldiers the necessary guidance and training focus to develop platoon and crew training strategies to achieve platoon collective tasks, battle drills, leader tasks, and individual task proficiency.

PLANNING TRAINING

1-46. Training is formally planned at company level and higher. Training plans use the collective tasks identified for training during the METL development process and the assessment of proficiency in those tasks, then translate them into training events based on the commander's visualized end state. Two types of training plans exist: long- and short-range.

1-47. Platoon leader's continuously assess the status (manning, equipping, and training) of the platoon during training and assist the commander in modifying the long range training plan to build cohesion and achieve required METL proficiency as they move through the ARFORGEN force pools. (Refer to ADP 7-0.)

TRAINING PRODUCTS

1-48. Platoon leaders and platoon sergeants provide input to the battalion commander to help determine a training strategy for their platoon and prepare training plans that enable the platoon to be ready within the ARFORGEN process. Platoon leaders assist the commander in developing training plans that enable them to attain proficiency in the tasks needed to conduct operations under conditions in the OE.

1-49. Platoon leaders have available to them training products they can use to assist them in developing training events on the tasks needed to be trained:

- **Individual and collective tasks.** These training products are linked together by how they are used in training the Soldier, individually and collectively. Individual tasks are performed by individual Soldiers, and can include leader tasks as well. Collective tasks are performed by crews, sections, or platoons in order to accomplish a mission or function.
- **Combined arms training strategies.** Combined arms training strategies (CATS) are strategies at the battalion level that contain platoon task selections.
- **Warfighter training support packages.** Warfighter training support packages (WTSPs) are assembled products and materials that units can use during training events to achieve proficiency on their METL.
- **Unit task lists.** Unit task lists (UTLs) are groupings of collective tasks a unit performs based on its table of organization and equipment (TOE).

INDIVIDUAL AND COLLECTIVE TASKS

1-50. Both individual and collective tasks are performed during platoon training to assess the proficiency of individuals and groups on their ability to perform the tasks to standard.

Individual Tasks

1-51. An individual task is a clearly defined, observable, and measurable activity accomplished by an individual. It is the lowest behavioral level in a job or duty that is performed for its own sake. An individual task supports one or more collective tasks or drills and often supports another individual task. Individual tasks can consist of both leader and staff tasks.

- **Leader tasks.** An individual task (skill level 2 or higher) a leader performs that is integral to the performance of a collective task.

- **Staff tasks.** A clearly defined and measurable activity or action performed by a staff (collective) or a staff member (individual) of an organization that supports a commander in the exercise of unit mission command.

Collective Tasks

1-52. A collective task is a clearly defined, observable, and measurable activity or action that requires organized team or unit performance, leading to the accomplishment of a mission or function. Collective task accomplishment requires the performance to standard of supporting individual or collective tasks. Collective tasks can consist of both shared and unique tasks.

- **Shared.** A shared collective task is a collective task that applies to or is performed by more than one type unit. Since the task, conditions, standards, task steps, and performance measures of shared collective tasks do not change, the collective task is trained and performed in the same way by all units that "share" the task. For example, Task 19-3-2406 (*Conduct Roadblock and Checkpoint*) can be conducted by Armor and Infantry companies.
- **Unique.** A unique collective task is a clearly defined, unit-specific collective task. For a collective task to be classified unique, no other unit or proponent (such as Infantry or engineers) may have the capability or requirement to perform the task. The designated proponent is solely responsible for the development and maintenance of a unique collective task.

1-53. These tasks are primarily performed in the operational domain, so the emphasis is on unit performance. Each collective task contains information that includes:

- **Assessment information.** Platoon leaders review the measures of performance and measures of success. They determine whether the platoon had performed those previously and what the assessment was when performed. If an assessment was conducted, this assessment can provide needed information advising if the platoon has performed the tasks and is considered trained (T), partially trained (P), or untrained (U).
- **General information.** These include task title and warfighting function.
- **Task data, conditions, and standards.**
- **Task attributes.** These are the tasks trained at night, under mission-oriented protective posture (MOPP) conditions, and the task steps.

- **Supporting information.** These are products/references, individual tasks, drills, and collective tasks, as well as the prerequisite collective tasks.

Unit Task Lists

1-54. The UTL is a product of mission analysis that identifies all of the collective tasks (shared and unique) that a unit is organized, manned and equipped to conduct. The UTL is produced for each unit with a TOE/Modified TOE (MTOE) or table of distribution and allowance (TDA).

1-55. The UTL can comprise existing collective tasks, or collective tasks identified to be designed and developed. The UTL provides the baseline for a unit CATS. A training developer creates the UTL by linking collective tasks to those missions identified on the TOE. This process ensures that units train the appropriate tasks to required proficiency levels.

UTL Locations

1-56. An assembled UTL is located in Appendix A of this TC. The UTL is also maintained and accessed within the Digital Training Management System (DTMS).

Digital Training Management System

1-57. The DTMS is a Web-based training management system that allows the unit to conduct mission and METL development; training planning and management; and track unit training by implementing the doctrine, tactics, techniques, and procedures outlined in ADP 7-0 and Training Management.

Combined Arms Training Strategy

1-58. The CATS is the Army's overarching strategy for the current and future training of the force. It describes how the Army trains the total force to standard in the institution, unit, and through self-development. It also identifies, quantifies, and justifies the training resources required to execute the training. Unit CATSs are built using the unit missions, the UTL, and are designed to reflect the METL.

1-59. Combined arms training strategies have replaced mission-training plans (MTPs). Combined arms training strategies provide a training path with recommendations of what and who to train. Combined arms training strategies support the platoon task selections within the battalion's METL training and are synchronized with ARFORGEN.

Types of CATSs

1-60. Combined arms training strategies are based on the unit's TOE mission (that support the METL), employment, capabilities, and functions. There are two types of CATSs: unit and functional.

Unit CATSs

1-61. Unit CATSs are TOE-based and unique to a unit type. Unit CATS development considers organizational structure, METL, and doctrine to organize the unit's collective tasks in a strategy that provides a path for achieving task proficiency.

1-62. A CATS comprises a menu of task selections that provide a base strategy for unit commanders to plan, prepare, and assess training to provide a flexible training strategy. Combined arms training strategies are also designed to train a capability with supporting training events and resources. The events are designed to be trained in a logical sequence, starting with the lowest echelon and adding echelons as the events get progressively more complex. The culminating, or run event, for a CATS is usually the highest level event designed to train and/or evaluate the entire unit.

1-63. Unit CATS provide commanders and platoon leaders a training strategy from which they develop their unit training plan to achieve collective task proficiency, as well as support the ARFORGEN phases. These strategies are flexible and are not intended to constrain leaders, but rather provide them with a menu of core mission/capabilities based training events. They provide leaders with a method to train all tasks that a unit is designed to perform and estimate the required resources to support event driven training. Unit CATS provide leaders with tools to plan, prepare for, and evaluate unit training.

Functional CATS

1-64. Functional CATSs address a functional capability common to multiple units and echelons. Functional CATS supplement Unit CATS. They may be based on missions or functions performed by most units that are not unique to a specific unit type, or they may be developed to train warfighting functions or operational themes that were not incorporated into unit CATS. Two examples of functional CATSs are mission command (currently listed in CATS as command and control), and force protection. Functional CATS contain most of the same data elements as unit CATS.

Task Selections

1-65. Task selections describe a specific capability/mission, and include collective tasks that support developing that capability. A frequency of training and the types of events that could be used to train the capability are also recommended.

1-66. Task selections are usually trained utilizing a series of crawl-walk-run events. Crawl-walk-run events provide options to accommodate training at the appropriate level of difficulty based on training readiness assessment. Each event provides recommendations for who and how to train, and resources required which support that training.

1-67. Platoon leaders can consolidate the necessary collective and individual tasks they determine are needed to be trained to support the mission-essential tasks (METs), which helps them determine the time, and resources needed to train these tasks to proficiency. A matrix showing the CATS task selections used within the mortar platoon that support the task groups of the higher unit's METL are located in Appendix C of this TC.

Training Events

1-68. Platoon leaders organize collective and individual tasks into standard Army training events that, when conducted, adhere to the principles of training mentioned earlier in this chapter.

1-69. Platoon leaders can also develop training events internally, such as classes and sergeants time training (STT), when using the crawl-walk-run training path. A platoon's progress through its training plan is based on time available and the platoon leaders and commander's assessment of task proficiency using the doctrinal process of assessing the training events.

CATS Locations

1-70. CATS is available online digitally from both DTMS and the Army Training Network (ATN). In a digital format, CATS provides numerous links to training materials, which can assist the commander and platoon leaders to develop the commander's plan and conduct training.

Warfighter Training Support Packages

1-71. Warfighter training support packages are complete, stand-alone, exportable training packages that integrate training products and materials needed to train one or more collective tasks and support critical individual tasks. Warfighter training support packages are task-based information packages that provide structured situational training scenarios for LVCG unit or institutional training.

1-72. Warfighter training support packages assist platoon leaders in training the collective tasks that support the battalion METL. Each WTSP includes materials to support planning, preparing, executing and assessing training exercises at respective echelons. The WTSP can aid the platoon leader throughout the UTM of the unit during various training exercises.

Warfighter Training Support Package Locations

1-73. Warfighter training support packages are exportable for use by the unit, unlike CATS and UTLs that can be accessed through DTMS or the ATN. Battalion WTSPs are located within the Maneuver Center of Excellence (MCoE) Collective Training Branch Website on Army Knowledge Online (AKO). To access this Website:

- Log into AKO.
- Copy and paste the Web address, (https://www.us.army.mil/suite/grouppage/130823), into the address bar.
- Select enter.
- Select desired WTSP.

LIVE, VIRTUAL, CONSTRUCTIVE, AND GAMING TRAINING

1-74. Platoon leaders can use LVCG training to enhance training, replicate battlefield conditions, balance resources, and sustain readiness. Unit leaders consider each to dictate the degree of simulation they plan for their unit during training events. Utilizing simulations within their platoon training enables platoon leaders to maximize many of the principles of training and to manage scarce resources.

LIVE

1-75. Live training is executed in field conditions using tactical equipment. It involves real people operating real systems.

1-76. Live training may be enhanced by training aids, devices, simulators, and simulations (TADSS) and tactical engagement simulation (TES) to simulate combat conditions. Use of simulation (CCMK) is used to adversely improve a unit marksmanship caliber. Initial Homestation Instrumented Training System (I-HITS) provides position location and weapons effects data for real-time exercise monitoring and AAR capability. Instrumentable-Multiple Integrated Laser Engagement System (I-MILES) has replaced basic Multiple Integrated Laser Engagement System (MILES) that is currently fielded. This system provides real-time casualty effects necessary for tactical engagements training in direct-fire, force-on-force training scenarios and instrumented training scenarios.

Note. No enhanced training can duplicate firing live rounds.

VIRTUAL

1-77. Virtual training is executed using computer-generated battlefields in simulators with the approximate characteristics of tactical weapon systems and vehicles. Virtual training is used to exercise motor control, decision making, and communication skills.

1-78. Sometimes called "human-in-the-loop" training, it involves real people operating simulated systems. Soldiers being trained practice the skills needed to operate actual equipment.

1-79. An example of virtual training is the close combat tactical trainer (CCTT). This system is equipped with the latest Force XXI Battle Command Brigade-and Below (FBCB2) in support of training the digital force. Dismounted Soldier is part of the CCTT program. It provides the capability to train Soldiers and units in all three elements of decisive action described in ADP 3-0..

CONSTRUCTIVE

1-80. Constructive training uses computer models and simulations to exercise command and staff functions. It involves simulated people operating simulated systems.

1-81. Constructive training can be conducted by units from platoon through echelons above corps. A command post (CP) exercise is an example of constructive training. Other examples are Multiuser Online Virtual Exercise (MOVE) and hands-on-trainer (HOT).

GAMING

1-82. Gaming is technology that employs commercial or government off-the-shelf, multigenre games in a realistic, semi-immersive environment to support education and training. The military uses gaming technologies to create capabilities to help train individuals and organizations. Gaming can enable individual, collective, and multiechelon training. Gaming can operate in a stand-alone environment or be integrated with live, virtual, or constructive enablers. It can also be used for individual education.

1-83. Employed in a realistic, semi-immersive environment, gaming can simulate operations and capabilities. Examples of gaming are Virtual Battlespace System 2 (VBS2), which is a fully interactive three-dimensional system that provides a premium synthetic environment suitable for military training. Other examples are DARWARS Ambush, which involves convoy operation training and platoon-level mounted infantry tactics and dismounted operations; urban simulation (URBANSIM); and Command Post of the Future (CPOF).

This page intentionally left blank.

Chapter 2
Crosswalks and Outlines

This chapter provides the mortar platoon leader two examples of a supporting collective task to the battalion's METL crosswalk (one for Infantry battalions [Table 2-1], the other for a combined arms battalion [Table 2-2]), along with the training and evaluation outlines (T&EOs). Each item can assist the platoon leader in developing training plans and conducting training.

SECTION I – SUPPORTING COLLECTIVE TASKS TO BATTALION METL CROSSWALKS

2-1. Platoon leaders focus their training efforts on training collective tasks that support the battalion METL. One of the many responsibilities of the platoon leader is to determine which tasks to train. These crosswalks are tools the platoon leaders can use as a starting point for selecting the supporting collective task to battalion METL. The supporting collective task to the battalion METL crosswalks (Table 2-2 and Table 2-3) are examples developed by the Directorate of Training and Doctrine (DOTD), MCoE.

Table 2-1. Primary references for conducting decisive actions

Decisive Action	References
Offensive, Defensive, and Security Operations	FM 3-90.5, *The Combined Arms Battalion* FM 3-22.90, *Mortars* ATTP 3-21.90, *Tactical Employment of Mortars.*
Stability Operations	FM 3-07, *Stability Operations.*
Defense Support of Civil Authorities	FM 3-28, *Civil Support Operations.*

Table 2-2. Example mortar platoon supporting collective tasks to Infantry battalion METL crosswalk

Mortar Platoon, Infantry Battalion		METs and Task Groups						
		Attack		Defend	Security		Stability	
Task Number	Task Title	Movement to Contact	Deliberate Attack	Area Defense	Screen	Area Security	Public Order & Safety	
07-2-9014	Occupy an Assembly Area (Platoon-Company)	X	X	X		X		
07-2-6063	Maintain Operations Security (Platoon-Company)	X	X	X	X	X		
07-2-1234	Conduct an Airborne Assault (Platoon-Company) (IBCT Only)	X	X					
07-2-1495	Conduct an Air Assault (Platoon-Company) (I & SBCT Only)	X	X					
07-2-5081	Conduct Troop-Leading Procedures (Platoon-Company)	X	X	X	X	X		
07-2-5027	Conduct Consolidation and Reorganization (Platoon-Company)	X	X	X		X		
07-2-1342	Conduct Tactical Movement (Platoon-Company)	X	X			X		
07-3-1351	Occupy a Mortar Firing Position (Section-Platoon)	X	X	X	X	X		
07-3-2045	Reconnoiter a Mortar Firing Position (Section-Platoon)	X	X	X	X	X		
07-3-3099	Fire a Mortar Priority Target Mission (Section-Platoon)	X	X	X	X	X		

Table 2-2. Example mortar platoon supporting collective tasks to Infantry battalion METL crosswalk (continued)

Mortar Platoon, Infantry Battalion		METs and Task Groups					
		Attack		Defend	Security		Stability
Task Number	Task Title	Movement to Contact	Deliberate Attack	Area Defense	Screen	Area Security	Public Order & Safety
07-3-9013	Conduct Action on Contact	X	X	X	X	X	
07-3-3054	Fire a Mortar Adjust Fire Mission (Section-Platoon)	X	X	X	X	X	
07-3-3072	Fire a Mortar Fire for Effect Mission (Section-Platoon)	X	X	X	X	X	
63-2-4546	Conduct Logistics Package (LOGPAC) Support			X	X		X
03-2-9224	Conduct Operational Decontamination	X	X	X		X	
08-2-0003	Treat Casualties	X	X	X	X	X	X
08-2-0004	Evacuate Casualties	X	X				

Table 2-3. Example mortar platoon supporting collective tasks to CAB METL crosswalk

Mortar Platoon, Combined Arms Battalion		METs and Task Groups						
		Attack		Defend	Security			Stability
Task Number	Task Title	Movement to Contact	Deliberate Attack	Area Defense	Screen	Guard	Area Security	Establish Public Order & Safety
07-2-9014	Occupy an Assembly Area (Platoon-Company)	X	X	X			X	
07-2-6063	Maintain Operations Security (Platoon-Company)	X	X	X	X	X	X	X
07-2-1234	Conduct an Airborne Assault (Platoon-Company) (IBCT Only)	X	X					
07-2-1495	Conduct an Air Assault (Platoon-Company) (I & SBCT Only)	X	X					
07-2-5081	Conduct Troop-leading Procedures (Platoon-Company)	X	X	X	X	X	X	X
07-2-5027	Conduct Consolidation and Reorganization (Platoon-Company)	X	X	X			X	
07-2-1342	Conduct Tactical Movement (Platoon-Company)	X	X				X	
07-3-1351	Occupy a Mortar Firing Position (Section-Platoon)	X	X	X		X		
07-3-2045	Reconnoiter a Mortar Firing Position (Section-Platoon)			X	X			

Table 2-3. Example mortar platoon supporting collective tasks to CAB METL crosswalk (continued)

Mortar Platoon, Combined Arms Battalion		METs and Task Groups							
		Attack		Defend	Security			Stability	
Task Number	Task Title	Movement to Contact	Deliberate Attack	Area Defense	Screen	Guard	Area Security	Establish Public Order & Safety	
07-3-3099	Fire a Mortar Priority Target Mission (Section-Platoon)	X	X	X					
07-3-9013	Conduct Action on Contact	X	X	X	X	X	X		
07-3-3054	Fire a Mortar Adjust Fire Mission (Section-Platoon)	X	X	X	X	X	X		
07-3-3072	Fire a Mortar Fire for Effect Mission (Section-Platoon)	X	X	X	X	X	X		
63-2-4546	Conduct Logistics Package (LOGPAC) Support			X			X	X	
03-2-9224	Conduct Operational Decontamination	X	X	X			X		
08-2-0003	Treat Casualties	X	X	X	X	X	X		
08-2-0004	Evacuate Casualties	X	X						

SECTION II – TRAINING AND EVALUATION OUTLINES

INTRODUCTION

2-2. This section provides the METL supporting collective tasks in the form of training and evaluation outlines. All T&EOs support unit missions, and individual T&EOs may support multiple missions within the decisive action operations.

2-3. Task outlines have multiple uses. Leaders and Soldiers within the unit can use T&EOs as a reference on how to perform a task. Commanders and leaders may use them to identify subordinate unit supporting tasks.

Observers or evaluators can use them to record and document the unit's task performance. The T&EOs can be used to train or evaluate a single task.

STRUCTURE

2-4. Each T&EO provides the task conditions and standards. They also provide a series of task steps and performance measures that serve as a logical guide for performing the task. The task steps are generally sequential, but they may be performed concurrently, or even omitted, based on METT-TC. The unit's ability to accomplish the task steps and performance measures indicates whether or not it is executing the task to standard. Table 2-2 lists METL tasks by METs and task groups, with task title and numbers to that specific T&EO.

FORMAT

2-5. Each T&EO displayed in this TC comprises the following:

- **Task**. This is a description of the action to be performed by the unit, and provides the task number.
- **References**. These are in parenthesis following the task number. The reference that contains the most information (primary reference) about the task is listed first.
- **Condition**. The condition is a written statement of the situation or environment in which the unit is to do the collective task.
- **Standard**. This states the performance criteria that a unit must achieve to successfully execute the task. This overall standard should be the focus of training and understood by every soldier. The trainer or evaluator determines the unit's training status using performance observation measurements (where applicable) and his judgment. The unit must be evaluated in the context of the METT-TC conditions. These conditions should be as similar as possible for all evaluated elements. This will establish a common base line for unit performance.
- **Task steps and performance measures**. This is a listing of actions that is required to complete the task. These actions are stated in terms of observable performance for evaluating training proficiency. The task steps are arranged sequentially along with supporting individual tasks and their reference. Leader tasks within each T&EO are indicated by an asterisk (*). Under each task step are listed the performance measures that must be accomplished to correctly perform the task step. If the unit fails to correctly perform one of these task steps to standard, it has failed to achieve the overall task standard.

- **GO/NO-GO column**. This column is provided for annotating the unit's performance of the task steps. When assessing training, evaluate each performance measure for a task step and place an "X" in the appropriate column. A major portion of the performance measures must be marked a "GO" for the task step to be successfully performed.
- **Supporting collective tasks**. An individual task is a clearly defined, observable, and measurable activity accomplished by an individual. It is the lowest behavioral level in a job or duty that is performed for its own sake.

USE

2-6. The T&EOs can be used to train or evaluate a single task.

TASK: Occupy an Assembly Area (Platoon-Company) (07-2-9014)

(FM 3-21.10)   (FM 3-21.8)

CONDITIONS: The unit conducts operations independently or as part of a higher headquarters (HQ) and receives an operation order (OPORD) or fragmentary order (FRAGO) to occupy an assembly area at the location and time specified. All necessary unit personnel and equipment are available. Indirect fire and close air support (CAS) are available. Unit has established communications with required headquarters/units at all echelons. Unit has received guidance on the rules of engagement (ROE). Coalition forces and noncombatants may be present in the operational environment. Civilians, government agencies, nongovernment organizations (NGOs), and local and international media may be in the area. Some iterations of this task should be performed under the mission variables; mission, enemy, terrain and weather, troops available, time available and civil considerations (METT-TC) that aid or limit performance. Some iterations of this task should be performed in mission-oriented protective posture 4 (MOPP 4).

STANDARDS: The unit occupies the assembly area according to the unit standing operating procedures (SOPs), the order, and/or higher commander's guidance. Unit enters the assembly area without stopping or blocking the route of march, moves all personnel and equipment to their assigned positions no later than (NLT) the time specified in the order. Unit establishes priorities of work, local security, and maintains appropriate readiness condition (REDCON) levels. Unit complies with the ROE, mission variables, mission instructions, higher HQ order, and other special orders.

TASK STEPS AND PERFORMANCE MEASURES	GO	NO-GO

PLAN
*1. Unit leaders receive an OPORD or a FRAGO that requires the unit to establish and an assembly area, and issue a warning order (WARNO) to the unit according to the troop-leading procedures (TLPs) and unit SOPs. The WARNO must include:
    a. Location and tentative timeline including movement times and NLT time for occupation of assembly area.
    b. Tentative unit organization for the operation, identifying special teams including security elements and quartering parties as required.
    c. Guidance directing the unit to conduct rehearsals on assigned tasks.		

TASK STEPS AND PERFORMANCE MEASURES	GO	NO-GO

*2. Unit leaders begin developing a tentative plan that includes the following actions:
 	
    a. Conducting mission analysis by using company intelligence support team (CoIST) focusing on the mission variables (METT-TC); taking advantage of maps, imagery, unmanned aircraft systems (UASs), unattended ground sensors (UGSs), and other available capabilities.
 	
    b. Developing a tentative course of action (COA), which includes:
 	
        (1) Identifying tentative rally points.
 	
        (2) Identifying likely enemy avenues of approach.
 	
        (3) Marking tentative dismount points on digital and conventional maps, as appropriate.
 	
        (4) Coordinating direct, indirect fire, CAS, and attack aviation, as appropriate.
 	
        (5) Developing task-organization; identifying personnel to perform in quartering party duties according to guidance and/or the unit SOPs.
 	
        (6) Developing contingency plans for actions on chance contact with the enemy.
 	
    c. Conducting composite risk management (CRM) to identify, assess, develop, and implement controls for hazards and to mitigate associated risks.
PREPARE
*3. Unit leaders issue the OPORD and use FRAGOs, as needed to redirect actions of subordinate elements.
4.   The unit conducts a rehearsal.
EXECUTE
5.   The quartering party moves to and clears the release point (RP), which includes the following actions:
 	
    a. Conducting reconnaissance of the route and the proposed assembly area.
 	
    b. Moving to and occupies assembly area.
 	
    c. Improving and marking entrances, exits, and internal routes.
 	
    d.   Conducting chemical, biological, radiological, and nuclear (CBRN) reconnaissance and monitoring.
 	
    e. Marking obstacles, mines, and contaminated areas.		

TASK STEPS AND PERFORMANCE MEASURES	GO	NO-GO
f. Selecting and marking tentative unit vehicle, weapons, and dismounted team positions according to OPORD, FRAGO, or unit SOPs.		
g. Developing a composite overlay of the assembly area.		
h. Maintaining surveillance and providing security of the area until the arrival of the unit.		
i. Posting guides with local security to guide unit to its initial position.		
6. The unit elements move to and clear the RP by taking the following actions:		
a. Moving to and occupies assembly area.		
b. Follows directions from guides and moves into marked positions.		
c. Orienting weapon systems to cover sectors of responsibility.		
d. Following proper cool-down procedures, shutting down engines simultaneously, if applicable.		
*7. Unit leaders or designated representatives initiate assembly area activities, which include the following actions:		
a. Reviewing organization of the assembly area with quartering party personnel.		
b. Assigning each element a sector of the perimeter to ensure mutual support and cover all gaps by observation and fire.		
c. Designating section direct fire responsibilities.		
d. Directing element leaders to prepare sector sketches.		
e. Keeping higher HQ informed of the status of the operation. This includes:		
(1) Reporting unit's arrival at the assembly area.		
(2) Reporting completion of initial occupation of assembly area positions.		
(3) Preparing and forwarding situation reports (SITREPs) to higher HQ, as necessary, throughout the operation.		
f. Determining security procedures, readiness condition, and priorities of work.		

TASK STEPS AND PERFORMANCE MEASURES	GO	NO-GO
8. The unit establishes and maintains local security, under directions from unit leaders by taking the following actions: a. Establishing patrols to prevent infiltration and to clear possible enemy observation posts (OPs) within assigned sector, if applicable. b. Designating OPs. c. Ensuring OPs have communications with the unit. d. Camouflaging equipment. e. Enforcing noise, light, and litter discipline. f. Preparing primary fighting positions. 9. The unit conducts the following tasks based on the priority of work established by the units leaders: a. Position key weapons systems, vehicles, and other assets to effectively cover each engagement area (EA). b. Conduct reconnaissance of the EA from the enemy's perspective, if possible. c. Assign sectors of fire, engagement priorities, and other fire control measures. d. Ensure the unit is tied in with the unit on its right and left. e. Designate final protective fires (FPFs) and final protective lines (FPLs). f. Clear fields of fire. g. Prepare range cards/sector sketches. h. Construct primary defensive positions in according to unit SOPs and/or as directed. i. Establish communications. j. Emplace claymore mines and protective obstacles, as required, by taking the following actions: (1) Identifying dead space and requirements to refine the location of the obstacle group and fire control measures. (2) Ensuring obstacles are covered by direct or indirect fire and under friendly observation. (3) Ensuring obstacles are concealed from enemy observation as much as possible. (4) Ensuring obstacles are employed according to the commanders' guidance.		

TASK STEPS AND PERFORMANCE MEASURES	GO	NO-GO
(5) Ensuring obstacles are tied in with existing obstacles. (6) The elements forward a sector sketch to the unit leaders and keep one for their use. (7) The unit performs field sanitation operations, which includes: a. Maintaining adequate supply of potable water. b. Establishing latrines and hand washing facilities. c. Performing personal hygiene activities. 12. The unit assumes a REDCON level according to unit SOPs or the order. The following are examples of REDCON levels: a. Maintain operations security a unit leader's guidance, order, and/or unit SOPs. b. Increases REDCON levels progressively, as required based on unit leader's guidance or unit SOPs. ASSESS *14. Unit leaders, or designated representatives, conduct preparations for departing the assembly area, on receiving order, by taking the following actions: a. Reconnoitering route and/or calculating time distance for departing the assembly area, as directed. b. Inspecting elements to ensure no equipment, supplies, or other items of tactical or intelligence value are left behind. c. Increasing REDCON levels progressively, as required, based on higher HQ's guidance or unit SOPs. *indicates a leader task step.		

SUPPORTING INDIVIDUAL TASKS

Task Number	Task Title
171-121-4045	Conduct Troop-Leading Procedures
171-610-0001	Perform a Map Reconnaissance
071-326-5770	Prepare a Platoon Sector Sketch
061-283-6003	Adjust Indirect Fire
301-348-1050	Report Information of Potential Intelligence Value

SUPPORTING INDIVIDUAL TASKS

Task Number	Task Title
071-326-5703	Construct Individual Fighting Positions
052-191-1362	Camouflage Equipment
061-284-3040	Engage Targets with Close Air Support

SUPPORTING COLLECTIVE TASKS

Task Number	Task Title
07-2-1342	Conduct Tactical Movement (Platoon-Company)
07-2-3027	Integrate Direct Fires (Platoon-Company)
07-2-3036	Integrate Indirect Fire Support (Platoon-Company)
07-2-5027	Conduct Consolidation and Reorganization (Platoon-Company)
07-2-6063	Maintain Operations Security (Platoon-Company)
07-2-9003	Conduct a Defense (Platoon-Company)
07-2-9005	Conduct a Linkup (Platoon-Company)
07-2-9007	Conduct a Passage of Lines as the Stationary Unit (Platoon-Company)
07-3-9016	Establish an Observation Post
08-2-0003	Treat Casualties
44-3-3220	Perform Passive Air Defense Measures
44-3-3221	Perform Active Air Defense Measures

SUPPORTING BATTLE/CREW DRILLS

Drill Number	Drill Title
07-3-D9501	React to Contact (Visual, IED, Direct Fire [includes RPG])
17-3-D8008	React to an Obstacle

TASK: Maintain Operations Security (Platoon-Company) (07-2-6063)

(FM 3-21.10)　　　(FM 3-21.8)

CONDITIONS: The unit conducts operations as part of a higher headquarters (HQ) and maintains operations security to deny the enemy information about friendly activities taking place in the area of operations (AO). All necessary personnel and equipment are available. The unit has communications with higher, adjacent, and subordinate elements. The enemy has the capability to detect the unit visually, audibly, and electronically. The unit has been provided guidance on the rules of engagement (ROE). Coalition forces and noncombatants may be present in the operational environment. Some iterations of this task should be conducted during limited visibility conditions. Some iterations of this task should be performed in mission-oriented protective posture 4 (MOPP 4).

STANDARDS: The unit maintains operations security according to the standing operating procedures (SOPs), the order, and or higher commander's guidance. The unit practices active and passive noise, light, litter, and communications discipline to deny friendly information to enemy forces. The unit complies with the ROE.

TASK STEPS AND PERFORMANCE MEASURES	GO	NO-GO
PLAN *1. Unit leaders gain and or maintain situational understanding using available communications equipment, maps, intelligence summaries, situation reports (SITREPs), and other available information sources. Intelligence sources include: company intelligence support team (CoIST), human intelligence (HUMINT), signal intelligence (SIGINT), and imagery intelligence (IMINT) to include unmanned aircraft systems (UASs) and unattended ground sensors (UGSs). PREPARE *2. The unit leader protects friendly information by checking or performing the following: 　a. Prohibits fraternization with civilians (as applicable). 　b. Makes sure maps do not contain the friendly order of battle. 　c. Conducts inspections and gives briefings to ensure that personnel do not carry details of military activities in personal materials (letters, diaries, notes, drawings, sketches, or photographs).		

TASK STEPS AND PERFORMANCE MEASURES	GO	NO-GO
d. Safeguards weapons, ammunition, sensitive items, and classified documents. e. Sanitizes all planning areas and positions before departure. EXECUTE 3. The unit employs active and passive security measures. It takes the following actions: a. Mans and performs surveillance from observation posts (OPs). b. Conducts counter reconnaissance patrols, if possible. c. Destroys enemy reconnaissance elements, when encountered. d. Conducts stand to procedures. e. Establishes local security. f. Establishes specific levels of alert based on the mission variables; mission, enemy, terrain and weather, troops and support available, time available, civil considerations (METT-TC). g. Uses camouflage. Takes the following actions: (1) Ensures fighting positions are camouflaged using exposed dirt to break the outline of a position. (2) Checks the position from the enemy's point of view. (3) Ensures camouflage nets (if used) are properly hung. (4) Employs overhead camouflage and sites positions and personnel to prevent detection from the air. (5) Keeps personnel and equipment in the shadows, when possible. (6) Ensures all personnel use camouflage paint to cover exposed skin. (7) Covers all reflective surfaces with non reflective material, such as cloth, mud, or camouflage stick. (8) Avoids crossing near footpaths, trails, and or roads, where possible. (9) Erases tracks leading into the positions. (10)Camouflages equipment by using vegetation to break up the outline of the equipment and covers all reflective surfaces.		

TASK STEPS AND PERFORMANCE MEASURES	GO	NO-GO
h. Enforces litter discipline. Takes the following actions:		
(1) Collects and back hauls trash during logistics runs during stationary periods.		
(2) Carries trash until it can be disposed of securely.		
i. Uses terrain and vegetation for concealment.		
j. Controls movement. Takes the following actions:		
(1) Uses the proper movement formation and movement technique.		
(2) Moves as fast as the situation permits.		
(3) Moves along covered and concealed routes.		
4. The unit practices radio discipline. It takes the following actions:		
a. Uses the proper radio procedures. Takes the following actions:		
(1) Changes frequencies and call signs according to signal operating instructions (SOIs) and or SOPs.		
(2) Uses varied transmission schedules and lengths.		
(3) Uses established formats to expedite transmissions.		
(4) Encodes messages or uses secure voice.		
(5) Uses brevity codes, when possible.		
(6) Uses the lowest power settings possible.		
(7) Avoids transmission patterns.		
(8) Maintains radio silence, as directed.		
b. Takes the following actions if jamming is suspected:		
(1) Continues to operate.		
(2) Disconnects the antenna.		
(3) Switches to the highest power.		
(4) Relocates the radio.		
(5) Uses directional antennas.		
(6) Turns the squelch off.		
c. Uses messenger and wire whenever possible, especially in static positions.		
d. Uses the radio in the quiet message mode. Transmits an arranged number of taps or uses the push-to-talk switch in the same manner.		

TASK STEPS AND PERFORMANCE MEASURES	GO	NO-GO

*5. The unit practices light discipline. It takes the following actions:
  a. Conceals lights necessary for planning or reading a map. (A poncho can be used for this purpose). Ensures all flashlights have filters.
  b. Removes or tapes personal items that may reflect light (identification tags, rings, and watches).
  c. Prohibits use of cigarettes and open fires during darkness or when the enemy can observe smoke and or fire.
  d. Uses only passive night observation equipment (if possible).
*6. The unit practices noise discipline. It takes the following actions:
  a. Identifies shortcomings in noise discipline during final inspections.
  b. Tapes down or removes items that make noise.
  c. Uses normal means of communication to the greatest extent possible.
  d. Prohibits talking except as required to conduct or plan operations.
  e. Ensures arm-and-hand signals are used, to the fullest extent possible, during daylight hours or with night vision devices.
ASSESS
7. Unit leaders conduct precombat checks (PCCs) and precombat inspections (PCIs).
8. Unit continues operations as directed.
* indicates a leader task step.		

SUPPORTING INDIVIDUAL TASKS

Task Number	Task Title
113-637-2001	Communicate via a Tactical Radio in a Secure Net
052-191-1362	Camouflage Equipment
052-191-1361	Camouflage Yourself and Your Individual Equipment
071-326-5705	Establish an Observation Post
071-720-0009	Conduct a Local Security Patrol
071-730-0006	Enforce Operations Security
071-331-0815	Practice Noise, Light, and Litter Discipline

SUPPORTING INDIVIDUAL TASKS

Task Number	Task Title
071-331-1000	Prepare a Platoon Early Warning System AN/TRS-2 for Operation
171-530-3013	Direct Tactical Employment of a Traffic Control Point

SUPPORTING COLLECTIVE TASKS

Task Number	Task Title
05-2-3002	Camouflage Vehicles and Equipment
07-2-6045	Employ Camouflage, Concealment, and Deception Techniques (Platoon-Company)
07-3-9016	Establish an Observation Post
07-3-9002	Conduct a Security Patrol

SUPPORTING BATTLE/CREW DRILLS

Drill Number	Drill Title
07-3-D9501	React to Contact (Visual, IED, Direct Fire [includes RPG])

TASK: Conduct an Airborne Assault (Platoon-Company) (07-2-1234

(FM 90-26)   (FM 3-21.10)

CONDITIONS: The unit conducts operations as part of a higher headquarters (HQ) and receives an operation order (OPORD) or fragmentary order (FRAGO) to conduct an airborne assault. Communications have been established, and information is being passed according to the unit standing operations procedures (SOPs). The unit has received guidance on the rules of engagement (ROE). Coalition forces and noncombatants may be present in the operational environment. Some iterations of this task should be conducted during limited visibility conditions. This task should not be trained in mission-oriented protective posture 4 (MOPP 4).

STANDARDS: The unit accomplishes the airborne assault according to SOPs, the order, and/or commander's guidance. The unit assembles its personnel and meets station and load times specified in the order. The unit assaults and secures the objective. The unit complies with the ROE.

TASK STEPS AND PERFORMANCE MEASURES	GO	NO-GO

PLAN
*1. Unit leaders gain and or maintain situational understanding using available communications equipment, maps, intelligence summaries; situation reports (SITREPs) and other available information sources. Intelligence sources include: company intelligence support team (CoIST), human intelligence (HUMINT), signal intelligence (SIGINT), and imagery intelligence (IMINT) to include unmanned aircraft systems (UASs) and unmanned ground systems (UGSs).
*2. The unit leader receives an OPORD or a fragmentary order FRAGO directing unit to conduct an airborne assault. The unit leader issues a warning order (WARNO) to element leaders ensuring that subordinate leaders have sufficient time for their own planning and preparation needs. The WARNO must include:
    a.   Tentative unit organization for the assault, identifying the security forces, main body, reserve, and sustaining organization, as applicable.
    b. Location and tentative timeline for the assault, including movement times and no later than time for execution.		

TASK STEPS AND PERFORMANCE MEASURES	GO	NO-GO
c. Guidance directing the unit to conduct rehearsals, initiate movement; initiate reconnaissance tasks and the commander's critical information requirements (CCIRs).		
*3. The unit leader develops a tentative plan according to troop-leading procedures (TLP). The leader takes the following actions:		
a. Conducts mission analysis focusing on the mission given, enemy forces and their capabilities, terrain and weather effects, troops available, time available to execute the operation, and civil considerations (METT-TC).		
b. Receives an updated report showing the location of forward and adjacent friendly elements, if applicable.		
c. Receives an updated enemy situational template for added fratricide prevention and increased force protection, if applicable.		
d. Clarifies priority intelligence requirement (PIR) requirements.		
e. Confirms any changes to the higher HQ and unit task or purpose.		
f. Confirms any changes to the scheme of maneuver.		
4. The unit leader conducts a map reconnaissance. He takes the following actions:		
a. Identifies marshalling area and or assembly area.		
b. Identifies the drop zone (DZ).		
c. Identifies routes to and or from the marshalling area, DZ, and objective.		
d. Identifies tentative security, support by fire, and assault positions for the objective, if applicable.		
e. Identifies likely enemy avenues of approach into the marshalling area, DZ, and objective.		
*5. The unit leader confirms/reviews the airborne assault plan from higher HQ. He takes the following actions:		
a. Confirms/reviews the marshalling plan.		
b. Confirms/reviews the air movement plan. The plan:		
(1) Includes the departure airfield/location for each serial.		
(2) Includes number and type of aircraft.		

TASK STEPS AND PERFORMANCE MEASURES	GO	NO-GO
(3) Includes load time(s).		
(4) Includes take off time(s).		
(5) Includes designated primary and alternate DZ.		
(6) Includes remarks such as special instructions, key equipment, and location of key element leaders.		
(7) Addresses tactical integrity.		
(8) Addresses cross leveling of leaders, key weapons, and key equipment.		
(9) Addresses self-sufficiency.		
NOTE: Each aircraft load should be self-sufficient so its personnel can operate effectively by themselves if other aircraft do not arrive at the DZ.		
(10) Includes an air movement planning work sheet.		
(11) Includes a bump plan.		
(12) Includes checkpoints.		
(13) Includes abort criteria.		
(14) Includes manifest.		
c. Confirms/reviews the landing plan. The plan includes:		
(1) Sequence of delivery.		
(2) Method of delivery.		
(3) Place of delivery (DZs, landing zone(s) [LZs], and/or extraction zone(s) [EZs]).		
(4) Time of delivery.		
(5) Assembly/reorganization plan.		
(6) Required reports to higher.		
(7) Security measures.		
(8) Designation of an assembly area.		
(9) Recovery of accompanying supplies, if applicable.		
(10) Reconnaissance efforts.		
(11) Coordination and final preparation before the attack, if applicable.		
(12) Time and/or conditions for element to move out on their mission.		
d. Confirms/reviews the ground tactical plan. Takes the following actions:		
(1) Confirms the assault objective and airhead line.		

TASK STEPS AND PERFORMANCE MEASURES	GO	NO-GO
(2) Determines reconnaissance and security element requirements.		
(3) Determines observation post(s) requirements.		
(4) Identifies or confirms boundaries.		
(5) Identifies task organization.		
(6) Determines resupply plan.		
e. Identifies direct fire responsibilities.		
f. Addresses actions on chance contact with the enemy.		
PREPARE		
*6. The unit leader disseminates reports, overlays, and other pertinent information to each element to keep them abreast of the situation.		
*7. The unit leader issues clear and concise tasking, orders and instructions to include ROE.		
8. The unit conducts a rehearsal.		
9. The unit moves to the assembly/marshalling area.		
10. The unit conducts assembly/marshalling area activities. It takes the following actions:		
a. Assembles, organizes, and marks personnel according to the OPORD.		
b. Ensures cross-loading of personnel.		
c. Conducts pre-jump training no earlier than 48 hours (24 hours preferred) to time on target.		
d. Rigs all equipment for combat.		
e. Conducts rehearsals according to the OPORD.		
f. Conducts final inspections.		
g. Moves not later than (NLT) the time specified in the OPORD.		
h. Inspects all personnel and equipment rigged by the jump master NLT the time specified in the OPORD.		
i. Meets aircraft load time.		
j. Meets Air Force station time.		
k. Receives safety briefing before takeoff.		
11. The unit conducts air movement under the jumpmaster's control. The jumpmaster:		
a. Maintains control over personnel and equipment.		
b. Maintains communication with subordinate elements.		
c. Remains oriented during flight.		

TASK STEPS AND PERFORMANCE MEASURES	GO	NO-GO

d. Executes appropriate warnings.
EXECUTE
12. The unit exits the aircraft on the order of the jump master.
13. The unit secures assault objective(s). It takes the following actions:
   a. Discards parachutes on the DZ. Reserves are placed on top of main parachute to prevent inflation.
   b. Removes all parachutes from runways and taxiways (if jumping on airfield) so as not to jeopardize follow-on operations.
   c. Establishes communications on all nets as required.
   d. Assembles personnel according to the assembly plan.
   e. Uses proper movement techniques.
   f. Recovers supplies and equipment.
   g. Treats and evacuates casualties.
   h. Sends a SITREP to the higher HQ commander.
i. Departs assembly area(s) to secure assault objectives within the time specified in the order.
ASSESS
14. The unit consolidates and reorganizes as needed.
15. The unit continues operations as directed.
*indicates a leader task step.		

SUPPORTING INDIVIDUAL TASKS

Task Number	Task Title
171-610-0001	Perform a Map Reconnaissance
071-326-5502	Issue a Fragmentary Order
071-326-5505	Issue an Operation Order at the Company, Platoon, or Squad Level
071-326-5503	Issue a Warning Order

SUPPORTING COLLECTIVE TASKS

Task Number	Task Title
07-2-5009	Conduct a Rehearsal (Platoon-Company)
07-2-5027	Conduct Consolidation and Reorganization (Platoon-Company)

SUPPORTING COLLECTIVE TASKS

Task Number **Task Title**
07-2-5063   Conduct Composite Risk Management (Platoon-Company)
07-2-6063   Maintain Operations Security (Platoon-Company)
07-2-9001   Conduct an Attack (Platoon-Company)
07-2-9014   Occupy an Assembly Area (Platoon-Company)
07-3-9013   Conduct Action on Contact
08-2-0003   Treat Casualties
08-2-0004   Evacuate Casualties

SUPPORTING BATTLE/CREW DRILLS

Drill Number **Drill Title**
07-3-D9501   React to Contact (Visual, IED, Direct Fire [includes RPG])

TASK: Conduct an Air Assault (Platoon-Company) (07-2-1495)

(FM 3-21.10)   (FM 3-21.8)   (ATTP 3-18.12)

CONDITIONS: The unit is conducting operations as part of a higher headquarters (HQ) and has received an operation order (OPORD) or fragmentary order (FRAGO) that requires it to perform an air assault mission. The pickup zone (PZ) and landing zone (LZ) have been designated in the order. Higher HQ coordinates aviation assets and conducts indirect fire support planning. All necessary personnel and equipment are available. The unit has communications with higher, adjacent, and subordinate elements. The unit has been provided guidance on the rules of engagement (ROE). Coalition forces and noncombatants may be present in the operational environment. Some iterations of this task should be conducted during limited visibility conditions. Some iterations of this task should be performed in mission-oriented protective measure 4 (MOPP 4).

STANDARDS: The unit performs the air assault according to the standing operating procedures (SOPs), the order, and/or higher commander's guidance. Unit briefs members on loading, safety, and unloading procedures. The unit arrives at the PZ and loads as per the loading plan by the specified time. The unit establishes and maintains security in their designated area of the PZ. The unit dismounts at the LZ, establishes security, and moves to designated area.

TASK STEPS AND PERFORMANCE MEASURES	GO	NO-GO
NOTE: Air assault planning and coordination is normally conducted at battalion or higher HQ. The battalion is the lowest level that has sufficient personnel to plan, coordinate, and control an air assault operation. PLAN *1. Unit leaders gain and/or maintain situational understanding using available communications equipment, maps, intelligence summaries, situation reports (SITREPs), and other available information sources. Intelligence sources include: company intelligence support team (CoIST), human intelligence (HUMINT), signal intelligence (SIGNIT), and imagery intelligence (IMINT) to include unmanned aircraft systems (UASs), and unattended ground sensors (UGSs). *2. Unit leaders receive an OPORD or FRAGO and issue a warning order (WARNO) to the unit.		

TASK STEPS AND PERFORMANCE MEASURES	GO	NO-GO

*3. Unit leaders confirm friendly and enemy situations. They takes the following actions:
    a. Receive an updated report showing the location of forward and adjacent friendly elements, if applicable.
    b. Receive an updated enemy situational template for added fratricide prevention and increased force protection, if applicable.
    c. Clarify priority intelligence requirement (PIR) requirements.
    d. Confirm any changes to the task organization and unit task or purpose.
    e. Confirm any changes to the scheme of maneuver.
*4. Unit leaders or designated representatives attend the air mission briefing, if possible.
*5. Unit leaders plan using troop-leading procedures. They takes the following actions:
    a. Conduct a map reconnaissance.
        (1) Identify pickup zone (PZ) and landing zone (LZ).
        (2) Identify tentative security positions, if required.
        (3) Identify likely enemy avenues of approach into the PZ and LZ.
        (4) Mark tentative dismount points on maps as appropriate.
    b. Identify direct fire responsibilities.
    c. Consider the enemy's capabilities, likely courses of action (COAs), and specific weapons capabilities.
    d. Address actions on chance contact with the enemy.
    e. Confirm procedures for calling for indirect fires.
    f. Select a straggler control point, if not assigned by higher HQ, for bumped personnel, if not provided by higher HQ.
    g. Coordinate to deconflict the fires of aircraft and troops on the LZ.
*6. Unit leaders disseminate reports, overlays, and other pertinent information to each element to keep them abreast of the situation.		

TASK STEPS AND PERFORMANCE MEASURES	GO	NO-GO
*7. Unit leaders organize the load and/or personnel based on the allowable cargo load as stated in the SOP, OPORD, and/or FRAGO. They take the following actions: a. Designate chalks and chalk leaders. b. Ensure tactical integrity is maintained (keeps teams together on the same aircraft). c. Ensure self-sufficiency is maintained (keeps weapon systems [for example, machine guns, Javelins, and their crews] on the same aircraft with ammunition and components). d. Ensure key personnel and weapons are distributed between the aircraft. e. Ensure each aircraft is loaded so that dismounting personnel can react promptly and contribute to mission accomplishment. *8. Unit leaders or a designated representative ensure unit personnel are briefed on the air assault operation. They take the following actions: a. Brief locations of LZ, PZ, and rendezvous points. b. Brief type of aircraft, if known. c. Brief allowable cargo load of aircraft. d. Brief priority of loading and bump plan. e. Ensure bumped personnel (if any) are told to report to the straggler control point. f. Brief contingency plan for downed aircraft (both for personnel on the aircraft and the remainder of the unit). g Brief landing plan that includes: (1) Explanation of how aircraft are landing on the LZ. (2) Aircraft off-load procedures. h. Coordination of fires of aircraft and troops on the LZ. *9. Unit leaders issue clear and concise taskings, orders, and instructions to include ROE. PREPARE 10. Unit conducts a rehearsal. *11.Unit leaders issue FRAGOs, as needed, to address changes to the plan identified during the rehearsal.		

TASK STEPS AND PERFORMANCE MEASURES	GO	NO-GO
*12. Unit leaders coordinate and/or synchronize actions of subordinate elements.		
*13. Unit leaders use FRAGOs as needed to redirect actions of subordinate elements.		
EXECUTE		
14. Unit moves to the LZ and prepares for loading according to the SOP. They take the following actions:		
a. Secure their assigned portion of the PZ while waiting to load the aircraft.		
b. Load at the time specified in the OPORD or FRAGO.		
c. Leave security with vehicles as needed.		
d. Enforce safety measures according to the SOP.		
15. Unit arrives and secures PZ according to the SOP or guidance. They take the following actions:		
a. Establish security to complete LZ activities and prevent surprise by the enemy.		
b. Establish blocking positions on all high speed or high threat avenues of approach into the LZ.		
c. Repulse enemy attacks and or counterattacks.		
d. Unit evacuates casualties on departing aircraft.		
e. Arrive at the object area with sufficient combat power to perform the specified mission.		
ASSESS		
16. Unit consolidates and reorganizes as needed.		
17. Unit continues operations as directed.		
* indicates a leader task step.		

SUPPORTING INDIVIDUAL TASKS

Task Number	Task Title
171-121-4045	Conduct Troop-Leading Procedures
071-326-5503	Issue a Warning Order
071-326-5505	Issue an Operation Order at the Company, Platoon, or Squad Level
071-326-5502	Issue a Fragmentary Order

SUPPORTING COLLECTIVE TASKS

Task Number	Task Title
07-2-1342	Conduct Tactical Movement (Platoon-Company)
07-2-5009	Conduct a Rehearsal (Platoon-Company)

07-2-5027	Conduct Consolidation and Reorganization (Platoon-Company)
07-2-5063	Conduct Composite Risk Management (Platoon-Company)
07-2-6063	Maintain Operations Security (Platoon-Company)
07-3-9013	Conduct Action on Contact
07-3-9017	Conduct Actions at Danger Areas
08-2-0004	Evacuate Casualties

SUPPORTING BATTLE/CREW DRILLS

Drill Number	Drill Title
17-3-D8008	React to an Obstacle
07-3-D9505	Break Contact
05-3-D0016	Conduct the 5C's

TASK: Conduct Troop-leading Procedures (Company/Platoon) (07-2-5081)

(ADP 5-0)   (FM 3-21.10)

CONDITIONS: The unit is conducting operations as part of a higher headquarters and has received a warning order (WARNO), an operation order (OPORD), or a fragmentary order (FRAGO) for an upcoming mission. Time is available to conduct troop-leading procedures (TLP). All necessary personnel and equipment are available. The unit has communications with higher, adjacent, and subordinate elements. The unit has been provided guidance on the rules of engagement (ROE). Coalition forces and noncombatants may be present in the operational environment. Some iterations of this task should be conducted during limited visibility conditions. Some iterations of this task should be performed in mission-oriented protective posture 4 (MOPP) 4.

STANDARDS: The unit conducts troop-leading procedures according to the standing operating procedures (SOPs) and appropriate field manual. All planning, coordination, preparations, rehearsals, inspections, and orders are conducted.

TASK STEPS AND PERFORMANCE MEASURES	GO	NO-GO
*1. Unit leaders gain and or maintain situational understanding using available communications equipment, maps, intelligence summaries, situation reports (SITREPs) and other available information sources. Intelligence sources include: company intelligence support team (CoIST), human intelligence (HUMINT), signal intelligence (SIGINT), and imagery intelligence (IMINT) to include unmanned aircraft systems (UASs) and unattended ground sensors (UGSs).		
2. Unit leader receives the mission.		
a. Conducts an initial assessment of the mission focused on mission, enemy, terrain and weather, troops and support available, time available and civil considerations (METT-TC) analysis.		
b. Identifies critical time requirements.		
*3. Unit leader issues WARNO to the unit that includes:		
a. Unit's mission.		
b. Units or elements participating in the operation.		
c. Time and place for issuing the OPORD.		
d. Specific tasks not addressed by unit SOP.		

TASK STEPS AND PERFORMANCE MEASURES	GO	NO-GO

e. Time line for the operation.
f. Movement instructions, if movement is to be initiated before OPORD issuance.
*4. Unit leader makes a tentative plan. (150-650-0094)
    a. Develops an estimate of the situation which includes. (150-650-0015, 150-650-0055)
       (1) Detailed mission analysis.
          (a) Understand the commander's intent and his concept.
          (b) Analyze METT-TC factors in as much detail as time and quality of information will allow.
          (c) Identify specified, implied and essential tasks.
          (d) Identify any constraints; for example, requirements for action and prohibition of actions.
          (e) Restate unit mission statement of who, what, when, where, and why.
       (2) Situational analysis and course of action (COA) development.
          (a) Determine one or more ways to achieve the mission.
          (b) Analyze relative combat power.
          (c) Generate options.
          (d) Array unit's initial forces.
          (e) Develop schemes of maneuver.
          (f) Assign headquarters.
       (3) Analysis of each COA (Wargame).
          (a) Determine advantages and disadvantages.
          (b) Visualize the flow of battle.
       (4) Comparison of each COA: Does the COA accomplish the unit purpose.
    b. Makes a decision on which COA will be used. (The decision represents the tentative plan.)
*5. Unit leader initiates movement according to the WARNO, or OPORD, or FRAGO and or unit tactical standing operating procedure (TSOP). (There may be a need to initiate movement immediately following the issuance of the WARNO.)
    a. Establishes movement control through the use of way points and or graphic control, measures.		

TASK STEPS AND PERFORMANCE MEASURES	GO	NO-GO
b. Assumes the appropriate readiness condition (REDCON) level.		
c. Dispatches quartering party as necessary.		
d. Begins priorities of work.		
e. Conducts time-distance check of the route to the start point (SP), as necessary.		
f. Prepares reconnaissance elements for movement.		
*6. Unit leader conducts reconnaissance.		
a. Coordinates with organizations that his reconnaissance elements will pass through or near.		
b. Completes a personal reconnaissance when time allows.		
c. Completes a map reconnaissance as a minimum when time does not allow a personal reconnaissance.		
d. Requests reconnaissance assets conduct the reconnaissance, if the risk of enemy contact is high.		
e. Focus on gaps identified in mission analysis and through wargaming.		
*7. Unit leader completes the plan.		
a. Adjusts the plan based on results of the reconnaissance.		
b. Ensures the plan meets the requirements of the mission and the commander's intent.		
c. Refines indirect fire target list.		
d. Prepares overlays.		
e. Coordinates sustainment requirements.		
*8. Unit leader issues order using the five paragraph order format.		
a. Ensures that all key leaders are presence.		
b. When possible issues the order from a vantage point overlooking the actual terrain the mission will be conducted on.		
c. Refers to terrain models, maps, overlays, Force XXI Battle Command Brigade and Below (FBCB2) displays and/or sketches to illustrate critical events or tasks.		
d. Conducts confirmation brief with subordinate leaders to ensure the following:		
(1) Subordinate leaders understand the enemy / threat and the friendly situations.		
(2) Subordinate leaders understand the commander's intent two levels up.		

TASK STEPS AND PERFORMANCE MEASURES	GO	NO-GO

(3) Subordinate leaders understand their assigned tasks.
*9. Unit leader supervises preparations and refines the plan.
   a. Conducts rehearsal that at a minimum includes: (07-2-5009)
      (1) Actions on the objective.
      (2) Actions on the enemy contact.
      (3) Backbriefs.
   b. Conducts inspections.
      (1) Conducts pre-combat inspections to ensure compliance with the OPORD and/or SOP.
      (2) Ensures task organization is according to OPORD and accepted by digital systems.
      (3) Confirms connectivity within the unit and between adjacent, supporting, and higher units/elements.
   c. Unit leader refines the plan.
      (1) Incorporates changes to correct deficiencies exposed during rehearsals.
      (2) Incorporates changes required by higher HQ FRAGO.
      (3) Incorporates changes based on information provided by ongoing reconnaissance and surveillance (R&S) operations.
      (4) Issues FRAGOs to subordinate units as required.
"*" indicates a leader task step.		

SUPPORTING INDIVIDUAL TASKS

Task Number	Task Title
171-121-4045	Conduct Troop Leading Procedures
071-326-5503	Issue a Warning Order
071-326-5505	Issue an Operation Order at the Company, Platoon, or Squad Level
071-326-5502	Issue a Fragmentary Order

SUPPORTING COLLECTIVE TASKS

Task Number	Task Title
07-2-5009	Conduct a Rehearsal (Platoon-Company)
07-2-5027	Conduct Consolidation and Reorganization (Platoon-Company)

SUPPORTING COLLECTIVE TASKS

Task Number	Task Title
07-2-5063	Conduct Composite Risk Management (Platoon-Company)
07-2-6063	Maintain Operations Security (Platoon-Company)

SUPPORTING BATTLE/CREW DRILLS

Drill Number	Drill Title
07-3-D9501	React to Contact (Visual, IED, Direct Fire [includes RPG])
07-3-D9504	React to Indirect Fire

TASK: Conduct Consolidation and Reorganization (Platoon-Company) (07-2-5027)

(FM 3-21.10)   (FM 3-21.8)

CONDITIONS: The unit conducts operations as part of a higher headquarters (HQ) and is in contact with the enemy. The unit must consolidate and reorganize. The unit communicates with higher, adjacent, and subordinate elements. Enemy forces have withdrawn to hasty defensive positions but have the capability to counterattack. The unit has guidance on the rules of engagement (ROE). Coalition forces and noncombatants may be present in the operational environment. Some iterations of this task should be performed in mission-oriented protected posture 4 (MOPP 4).

STANDARDS: The unit consolidates and reorganizes according to the standing operating procedures (SOPs) and/or higher commander's guidance. The unit occupies a hasty fighting position with sectors of fire, establishes security, accounts for all personnel and equipment, and reestablishes the chain of command. Wounded in action (WIAs) are identified, stabilized, and prepared for evacuation. Killed in action (KIAs) are identified and prepared for evacuation. Ammunition and supplies are redistributed as needed.

TASK STEPS AND PERFORMANCE MEASURES	GO	NO-GO
PLAN		
*1. Unit leaders gain and/or maintain situational understanding using available communications equipment, maps, intelligence summaries, situation reports (SITREPs), and other available information sources. Intelligence sources include: company intelligence support team (CoIST), human intelligence (HUMINT), signal intelligence (SIGINT), and imagery intelligence (IMINT) to include unmanned aircraft systems (UASs) and unattended ground sensors (UGSs).		
*2. Unit leaders confirm friendly and enemy situations. They receive an updated:		
a. Report showing the location of forward and adjacent friendly elements.		
b. Enemy situational template for added fratricide prevention and increased force protection.		
*3. Unit leaders conduct troop-leading procedures.		
PREPARE		

TASK STEPS AND PERFORMANCE MEASURES	GO	NO-GO

*4. Unit leaders position the observation post (OP) forward to provide security. They ensure that:
    a. Members are alert for a possible counterattack.
    b. The unit main body is not engaged without warning.
EXECUTE
5. The unit occupies hasty fighting positions near the objective. It takes the following actions:
    a. Establishes local security, including:
        (1) Evaluates terrain thoroughly.
        (2) Positions the elements using the clock or the terrain feature technique.
        (3) Mans key weapons, as required by factors of mission, enemy, terrain and weather, troops and support available, time available, civil considerations (METT-TC).
    b. Destroys all organized resistance.
    c. Conducts reconnaissance of objective and/or area of operations (AO) to ensure it is free of enemy.
    d. Defends against enemy counterattacks.
    e. Begins decontamination operations, if required and as factors of METT-TC dictate.
    f. Establishes the chain of command.
    g. Establishes communications.
*6. Unit leaders assign elements temporary sectors of fire.
*7. Unit leaders ensure subordinate leaders adjust positions to cover likely avenues of approach and ensure mutual support between elements and adjacent units.
*8. The unit secures enemy prisoners of war (EPWs).
*9. Unit leaders report intelligence information of immediate value to next higher HQ.
*10. Unit leaders supervise redistribution of ammunition and equipment.
*11. Unit leaders provide ammunition, casualty, and equipment (ACE) reports to the headquarters.
*12. Unit leaders coordinate resupply.
*13. The unit treats and evacuates casualties.
*14. The unit processes captured documents and/or equipment as required.
ASSESS		

TASK STEPS AND PERFORMANCE MEASURES	GO	NO-GO
*15. The unit continues operations as directed. *indicates a leader task step		

SUPPORTING INDIVIDUAL TASKS

Task Number	Task Title
171-121-4045	Conduct Troop-Leading Procedures
171-121-4038	Supervise Local Security
031-507-3014	Supervise Decontamination Procedures
113-571-1022	Perform Voice Communications
081-831-1058	Supervise Casualty Treatment and Evacuation
071-940-0002	Conduct Resupply of a Platoon
301-371-1200	Process Captured Materiel

SUPPORTING COLLECTIVE TASKS

Task Number	Task Title
07-2-6063	Maintain Operations Security (Platoon-Company)
07-3-9016	Establish an Observation Post
08-2-0003	Treat Casualties
08-2-0004	Evacuate Casualties
19-3-3107	Process Detainee(s) at Point of Capture (POC)

SUPPORTING BATTLE/CREW DRILLS

Drill Number	Drill Title
05-3-D0016	Conduct the 5C's
07-3-D9507	Evacuate a Casualty (Dismounted and Mounted)

TASK: Conduct Tactical Movement (Platoon-Company) (07-2-1342)

(FM 3-21.10)   (FM 3-21.8)

CONDITIONS: The unit conducts operations as part of a higher headquarters (HQ) and receives an operation order (OPORD) or fragmentary order (FRAGO) that requires the unit to conduct a tactical movement. The unit must move tactically to prevent the enemy from detecting its activities or intent. All necessary personnel and equipment are available. The unit has communications with higher, adjacent, and subordinate elements. The unit has been provided guidance on the rules of engagement (ROE). Coalition forces and noncombatants may be present in the operational environment. Some iterations of this task should be conducted during limited visibility conditions. Some iterations of this task should be performed in mission-oriented protective posture 4 (MOPP 4).

STANDARDS: The unit conducts tactical movement according to the standing operating procedures (SOPs), the order, and/or the commander's guidance. The unit moves using the route or axis of advance, formations, and techniques of movement as specified or as dictated by factors of mission, enemy, terrain and weather, troops and support available, time available, and civil considerations (METT-TC). The unit maintains the appropriate interval and all-round security during movement.

TASK STEPS AND PERFORMANCE MEASURES	GO	NO-GO

PLAN
*1. Unit leaders gain and or maintain situational understanding using available communications equipment, maps, intelligence summaries, situation reports (SITREPs), and other available information sources. Intelligence sources include: company intelligence support team (CoIST), human intelligence (HUMINT), signal intelligence (SIGINT), and imagery intelligence (IMINT) to include unmanned aircraft systems (UASs) and unmanned ground systems (UGSs).
*2. The unit leader receives an OPORD or FRAGO and issues a warning order (WARNO) to the unit. The WARNO must include:
    a. Tentative timeline for the tactical movement.
    b. Tentative unit organization for the tactical movement.
    c. Guidance directing the unit to conduct rehearsals.		

TASK STEPS AND PERFORMANCE MEASURES	GO	NO-GO

*3. The unit leader plans using troop-leading procedures (TLP). It takes the following actions:
    a. Conducts analysis based on factors of METT-TC.
    b. Considers the enemy's capabilities, likely courses of action (COAs), and specific weapons capabilities.
    c. Conducts a map reconnaissance. Takes the following actions:
        (1) Identifies routes that provide protection from direct and indirect fires. Takes the following actions:
            (a) Offers concealment from ground and air.
            (b) Avoids sky lining.
            (c) Avoids moving directly forward from firing positions.
            (d) Avoids danger areas and potential kill zones.
            (e) Avoids obvious enemy avenues of approach.
        (2) Identifies coordination points, passage points, and boundaries.
        (3) Identifies adjacent units.
        (4) Identifies potential danger areas.
        (5) Plots way points on easily recognizable terrain and on significant turns on the route for ease in navigation.
        (6) Marks tentative dismount points on maps as appropriate.
    d. Develops control measures. Takes the following actions:
        (1) Develops limited visibility marking to aid in mission command at night.
        (2) Develops graphics, if not assigned by higher HQ.
        (3) Designates guides, if possible.
        (4) Plans to use night vision devices and thermal devices, if available.
    e. Develops a movement plan. Takes the following actions:
        (1) Selects a tentative formation based on factors of METT-TC (mounted and or dismounted).		

TASK STEPS AND PERFORMANCE MEASURES	GO	NO-GO
(2) Coordinates formation with other elements moving in the main body formation, if using a different formation than the remainder of the higher HQ.		
(3) Selects the tentative movement technique based on factors of METT-TC.		
f. Plans to employ dismounted elements, if mounted, when any of the following conditions apply:		
(1) Detailed reconnaissance is required.		
(2) Stealth is required.		
(3) Enemy contact is expected or visual contact has been made.		
(4) Vehicle movement is restricted by terrain.		
(5) Time is not limited.		
(6) Security is the primary concern.		
g. Coordinates linkup with vehicles, if dismounted.		
h. Integrates indirect fire support for mounted and dismounted tactical movement.		
i. Identifies direct fire responsibilities.		
j. Organizes the unit as necessary to accomplish the mission and or compensate for combat losses.		
k. Develops a security plan.		
l. Assigns each element a sector of responsibility.		
m. Coordinates passage of lines as needed.		
n. Addresses actions on chance contact with the enemy.		
PREPARE		
*4. The unit leader disseminates reports, overlays, and other pertinent information to subordinates to keep them abreast of the situation.		
*5. The unit leader briefs the movement plan. He takes the following actions:		
a. Specifies conditions under which the unit will change movement techniques and or formations.		
b. Designates bounding and overwatch elements based on METT-TC.		
c. Briefs way points for checkpoints, boundaries, and so forth.		
d. Briefs commands used for mission command.		
*6. The unit leader issues clear and concise tasking, orders and instructions to include rules of engagement (ROE).		

TASK STEPS AND PERFORMANCE MEASURES	GO	NO-GO
7. The unit conducts a rehearsal.		
*8. The unit leader and reconnaissance element conducts the reconnaissance (based on METT-TC).		
*9. The unit leader adjusts the plan based on updated intelligence and reconnaissance effort.		
*10. The unit leader disseminates updated reports, overlays, and other pertinent information.		
EXECUTE		
*11. The unit leader or designated representative initiates movement to the line of departure (LD).		
*12. The unit leader coordinates/synchronizes actions of subordinate elements.		
*13. The unit leader uses FRAGOs as needed to redirect actions of subordinate elements.		
14. The unit moves using the following fundamentals:		
a. Moves as squads, section, and platoons.		
b. Moves as fast as the situation allows.		
c. Makes contact with the smallest element possible.		
d. Positions unit's key weapons and elements so they can provide responsive fires in the event of enemy contact.		
15. The unit moves using the appropriate formation as designated by the unit leader. It takes the following actions:		
a. Selects the formation that provides the proper control, security, and speed.		
b. Adjusts formation during limited visibility to maintain visibility between vehicles, individuals, teams, and squads, and to maintain the rate of movement.		
16. The unit executes movement technique as directed by the unit leader. It takes the following actions:		
a. Adjusts the movement technique to provide greater security, as the probability of enemy contact increases.		
b. Employs traveling technique as ordered. Takes the following actions:		
(1) Assumes correct order of march as directed by the unit leader.		
(2) Moves in a unit column with 20 to 50 meters between elements.		

TASK STEPS AND PERFORMANCE MEASURES	GO	NO-GO
NOTE: Distance between elements depends on visibility afforded by terrain, weather, and light.		
(3) Uses terrain driving techniques by individual elements to reduce exposure, if applicable.		
(4) Maintains visual contact between lead and trail elements.		
(5) Maintains unit integrity as much as possible and organizes the formation for ease of deployment during the upcoming mission.		
(6) Maintains areas of responsibility for observation and fire to ensure 360-degree security.		
c. Employs traveling overwatch technique as ordered. Takes the following actions:		
(1) Moves in a unit column or wedge.		
NOTE: Distance between elements is not fixed.		
(2) Positions an element as lead and elements as trail:		
(a) Lead element moves continuously.		
(b) Trailing element stays far enough behind the lead element to avoid fire directed at the lead element.		
(c) Trailing elements stay close enough so they can provide fire support or maneuver when the lead element makes contact.		
(d) Trailing elements halt periodically at advantageous vantage points to provide overwatch and or base of fire for the lead element.		
(3) Ensures vehicles use terrain driving techniques to reduce exposure.		
(4) Maintains command and control of the elements.		
(5) Maintains 360-degree security.		
d. Employs bounding overwatch technique as ordered.		
NOTE: The unit leader designates the initial bounding and initial overwatch elements and specifies either the alternate or successive bounding method. He takes the following actions:		
(1) Employs alternate bounds. Takes the following actions:		

TASK STEPS AND PERFORMANCE MEASURES	GO	NO-GO
(a) Ensures trailing element advances past the lead element to the next overwatch position. (This method is usually more rapid than successive bounds.) (b) Repeats sequence of bounding past each other until the unit halts, the movement technique is changed, or the unit transitions to maneuver by conducting actions on contact. (2) Employs successive bounds. Takes the following actions: (a) Ensures trailing element moves to an overwatch position that is approximately abreast of the lead element. (This method is easier to control and more secure than alternate bounding, but it is slower.) (b) Repeats sequence of bounding abreast each other until the unit halts, the movement technique is changed, or the unit transitions to maneuver by conducting actions on contact. *17. The unit leader positions himself where he can best control and execute the desired formation. 18. The unit maintains formation with correct interval, speed, and or lateral dispersion according to the unit leader's guidance or SOPs. 19. The unit orients weapon systems to provide security and maximize firepower as needed. 20. The unit moves undetected to the designated point specified in the OPORD. ASSESS 21. The unit consolidates and reorganizes as needed during and or after movement. 22. The unit continues operations as directed. * indicates a leader task step.		

SUPPORTING INDIVIDUAL TASKS

Task Number	Task Title
171-610-0001	Perform a Map Reconnaissance
071-326-5610	Conduct Movement Techniques by a Squad
071-410-0010	Conduct a Leader's Reconnaissance
052-192-3262	Prepare for an Improvised Explosive Device (IED) Threat Prior to Movement (Unclassified/For Official Use Only) (U//FOUO)

SUPPORTING INDIVIDUAL TASKS

Task Number	Task Title
052-703-9107	Plan for an Improvised Explosive Device (IED) Threat in a COIN Environment (Unclassified/For Official Use Only) (U//FOUO)
071-326-5502	Issue a Fragmentary Order
071-326-5503	Issue a Warning Order
071-326-5630	Conduct Movement Techniques by a Platoon

SUPPORTING COLLECTIVE TASKS

Task Number	Task Title
07-2-3000	Conduct Support by Fire (Platoon-Company)
07-2-3027	Integrate Direct Fires (Platoon-Company)
07-2-3036	Integrate Indirect Fire Support (Platoon-Company)
07-2-5009	Conduct a Rehearsal (Platoon-Company)
07-2-5027	Conduct Consolidation and Reorganization (Platoon-Company)
07-2-5063	Conduct Composite Risk Management (Platoon-Company)
07-2-6063	Maintain Operations Security (Platoon-Company)
07-2-9006	Conduct a Passage of Lines as the Passing Unit (Platoon-Company)
07-3-9013	Conduct Action on Contact
07-3-9017	Conduct Actions at Danger Areas
08-2-0004	Evacuate Casualties
44-3-3220	Perform Passive Air Defense Measures
44-3-3221	Perform Active Air Defense Measures

SUPPORTING BATTLE/CREW DRILLS

Drill Number	Drill Title
07-3-D9501	React to Contact (Visual, IED, Direct Fire [includes RPG])
07-4-D9316	Perform Direct Alignment of a Mortar

TASK: Occupy a Mortar Firing Position (Platoon/Squad) (07-3-1351) (FM 3-22.90) (FM 3-21.8)

CONDITIONS: The mortar platoon is conducting operations as part of a larger force and has received an operation or fragmentary order (OPORD or FRAGO) to occupy a firing position at a specified time. Latest intelligence indicates that a platoon-sized enemy element is withdrawing to establish defensive positions. The platoon has all necessary personnel and equipment including, if available, the Mortar Fire-Control System (MFCS); communications with higher, adjacent, supporting, and subordinate elements; and guidance on the rules of engagement (ROE). If an advance party is available, then the position is already prepared when the platoon arrives, and vice versa. The operational environment might include coalition forces or noncombatants. Some iterations of this task should be performed in limited visibility. Some iterations of this task should be performed in mission-oriented protective posture 4 (MOPP 4).

STANDARDS: The mortar platoon selects a firing position with good cover, concealment, natural lines of drift, multiple withdrawal routes, solid ground, and overhead and mask clearance; and that is located well away from human habitation (depending on the situation) and from known or suspected enemy locations. The platoon moves all personnel and equipment into their assigned positions no later than the time specified in the order. Within two minutes after occupying the mortar position, the mortar platoon was prepares to provide indirect-fire support to the company or battalion. The platoon is not surprised by the enemy.

If the mortar platoon sent an advance party to prepare the firing position, then the platoon completes the aiming circle before the weapon system carriers (81- and 120-mm mortars only) began to arrive. Timing begins when the first mortar carrier halted in position.

If the mortar platoon did not send an advance party, then the platoon prepares the position themselves and did not complete the aiming circle in advance. Timing begins either when the mortar platoon placed the aiming circle or M2 compass on the ground, or when the first mortar carrier halted in position, whichever occurred first. The mortar platoon lay and emplaces all mortars and aiming posts within the time standards in Table 2-3 (without the MFCS) or Table 2-4 (with MFCS).

The unit complies with the ROE.

Table 2-3. Time standards without MFCS

Number of Mortars	Reciprocal Lay		Placement of Aiming Post	
	Day	Night	Day	Night
One	2 min 15 sec	4 min	1 min, 15 sec	3 min
Two	3 min 15 sec	5 min	2 min	3 min
Three	4 min 15 sec	7 min	2 min, 30 sec	4 min
Four	5 min 15 sec	8 min, 30 sec	3 min	4 min
Five	6 min 15 sec	10 min	3 min, 30 sec	5 min
Six	7 min 15 sec	11 min, 30 sec	4 min	5 min

Table 2-4. Time standards with MFCS

Section Laid and Ready to Fire	
Day	Night
60 sec	1 min, 30 sec

TASK STEPS AND PERFORMANCE MEASURES	GO	NO-GO
*1. Platoon leader gain and or maintain situational understanding using available communications equipment, maps, intelligence summaries, situation reports (SITREPs) and other available information sources. Intelligence sources include: company intelligence support team (CoIST), human intelligence (HUMINT), signal intelligence (SIGNIT), and imagery intelligence (IMINT) to include unmanned aircraft systems (UASs) and unattended ground sensors (UGSs). **NOTE:** The MFCS is an automated fire-control system that integrates mortar platoons into current and future fire-support command architecture. The MFCS allows mortar squads to operate semiautonomously. This gives them some flexibility in tactical employment, even though they remain under the control of a central fire-direction center (FDC). *2. Leader conducts troop-leading procedures. *3. The mortar element senior leader plans and selects both a tentative and an alternate mortar firing position.		

TASK STEPS AND PERFORMANCE MEASURES	GO	NO-GO
NOTE: Position navigation aids (POSNAV) supplement, but do not replace, basic navigational skills. POSNAV allow leaders to track their own locations and those of higher, adjacent, and subordinate units. POSNAV provide directional information for movement and target acquisition and augment operational planning graphics (such as checkpoints, boundaries, coordination points, and phase lines).		

4. The mortar element moves to occupy the position.
   a. Moves on covered and concealed routes.
   b. Avoids likely ambush sites and danger areas.
   c. Maintains operations security.
   d. Maintains all-round security.
*5. The mortar element senior leader--
   a. Halts the element 200 to 400 meters from the tentative mortar firing position.
   b. Reconnoiters and confirms the location of the position.
   c. Designates the point of entry as six o'clock.
   d. Designates the center of the base as the command post (CP) and, upon arrival there, designated the actual mortar firing positions.
*6. Subordinate element leaders reconnoiter their assigned positions.
7. The mortar element occupies the mortar firing position. Takes the following actions:
   a. [Each subordinate element leader] Upon arriving in his firing position, designates the mortar firing positions.
   b. [Each gunner] Orients the mortar in the designated direction of fire. If the terrain permitted, he establishes intervals between mortars of 25 to 30 meters for the 60mm; 35 to 45 meters for the 81mm; and 60 meters for the 120mm (60 meters distance between positions not required with MFCS).
   c. [Each subordinate element leader] Ensures that the vehicles were positioned in defilade where natural camouflage concealed them.
   d. [Each subordinate element] Performs Crew Drill, "Place Mortar Into Action."

TASK STEPS AND PERFORMANCE MEASURES	GO	NO-GO
e. [Element leader] Positions the FDC near the middle of the firing positions so that the gunners could hear the FDC.		

8. The mortar element establishes all-round local security. Takes the following actions:

a. (Leader.) Positions observation posts (OPs) to observe likely enemy avenues of approach, and to provide early warning to protect the main body from surprise. (07-3-9016)

b. Gunners.) Orients on likely enemy avenues of approach.

9. The mortar element lay the mortars.

10. Unless the advance party has already done this, the aiming circle operator correctly mounts, levels, and orients the aiming circle (81- and 120-mm mortars only).

11. Element(s) performs one or more of the following:

a. Crew Drill, "Reciprocal Lay With Aiming Circle."

b. Crew Drill, "Reciprocal Lay Using the Mortar Sight."

12. The FDC prepares to receive calls for fire.

a. Prepares the mortar ballistic computer (MBC) or plotting board.

b. Completes the firing chart within five minutes of arrival, unless the FDC initialized the MBC for firing.

NOTE: Digital enhancements change few indirect-fire procedures, but they do streamline the planning and calling of indirect fire. The ability to share information between echelons eases fire planning at all levels. (For example, a platoon leader can get the fire-support overlay fast using the V2/3 computer module and the Dismounted Soldier-System Unit (DSSU).) This lets leaders and soldiers call for fire digitally. Then, the digital call for fire can go to any indirect-fire asset on the net, but it normally routes through the Advanced Field Artillery Tactical Data System (AFATDS).

NOTE: Timing for the FDC began simultaneously for the laying of the mortars.

*13. Each subordinate element leader obtains an auxiliary aiming point for his squad's mortar. (071-074-0008)

a. Choose an auxiliary aiming point.

TASK STEPS AND PERFORMANCE MEASURES	GO	NO-GO
b. [Gunner] Obtains the deflection to the auxiliary aiming point.		
c. Records deflection to the auxiliary aiming point.		
14. The mortar element establishes internal wire communications. Takes the following actions:		
a. Ensures that every element had field telephone communications from its mortar position to the FDC.		
b. Within 10 minutes after aiming posts were emplaced, ensures that the field telephones were operational between the FDC and all elements.		
15. The mortar element constructs a mortar firing position. Takes the following actions:		
a. Marks the orientation of the firing position with luminous or reflective ground stakes.		
b. Establishes temporary mortar position either to the front or rear of the site where the unit was to construct the position.		
c. Camouflages the position to blend with the surrounding terrain, using natural vegetation when possible.		
d. Constructs the combat position.		
(1) Width -- three M16 rifle lengths across at the top, with the sides sloping slightly in toward the bottom.		
(2) Depth --18 to 20 inches (one-half the length of an M16) below ground level.		
e. Constructs a parapet around the dug-out part of the position.		
(1) Width -- 18 to 20 inches (one or two M16 lengths).		
(2) Openings -- one for the entrance to the position, and one each to the front and the rear, for sighting on the aiming posts.		
NOTE: To re-lay the mortar after completing the parapet, the mortar platoon removed a part of the parapet. This allowed the aiming-circle operator to see the mortar sight. The unit constructed the firing position for the 120-mm mortar using the same procedure and dimensions as for the 60 and 81-mm mortars, except that the 120-mm's position should measure three-and-a-half M16 rifle lengths wide.		
16. The mortar unit improves the firing position. Takes the following actions:		

TASK STEPS AND PERFORMANCE MEASURES	GO	NO-GO
a. (Leader.) Develops a plan for unit defense and supervised its preparation.		
b. Build primary mortar positions.		
c. (Leader.) Effectively integrates all available direct-fire weapons into the unit's perimeter defense.		
d. (Leader.) Plans indirect fires on avenues of approach.		
e. Selectively clears fields of fire.		
f. (Leader.) Designates alternate mortar positions and confirms supplementary ones.		
g. Rehearses occupation of alternate and supplementary individual fighting positions.		
h. If time permitted, build alternate and supplementary mortar positions.		
i. Emplaces early-warning devices.		
j. (Leader.) Develops obstacle plan. If time permitted, emplaced the obstacles.		
k. (Leader.) Prepares sector sketches and marked--		
(1) Observation posts and patrol routes, if any.		
(2) Maximum engagement lines (MELs) for primary weapon systems.		
(3) Mines and obstacles.		
(4) Squad positions (primary, alternate, and supplementary) and sectors of fire.		
(5) Direction of north.		
(6) Unit designation up to company level.		
(7) Date-time group.		
(8) Unit CP.		
l. (Leader.) Forwards the sector sketch to the company commander and kept a copy.		
NOTE: See task "Maintain Operations Security" for subtasks and standards related to camouflage.		
17. The platoon complies with ROE.		
* indicates a leader task step.		

SUPPORTING INDIVIDUAL TASKS

Task Number	Task Title
071-074-0012	Conduct Occupation of a Mortar Firing Position by a Squad
171-610-0001	Perform a Map Reconnaissance

SUPPORTING COLLECTIVE TASKS

Task Number	Task Title
07-2-1342	Conduct Tactical Movement (Platoon-Company)
07-2-5063	Conduct Composite Risk Management (Platoon-Company)
07-2-6063	Maintain Operations Security (Platoon-Company)
07-3-9016	Establish an Observation Post

SUPPORTING BATTLE/CREW DRILLS

Drill Number	Drill Title
07-4-D9267	Place 81-mm Mortar into Action
07-4-D9280	Mount the Mortar Carrier

TASK: Reconnoiter a Mortar Firing Position (Platoon/Squad) (07-3-2045) (FM 3-22.90)

CONDITIONS: The mortar platoon has received an order to occupy a new firing position, and it has just enough time to conduct a ground reconnaissance. Latest intelligence indicates that the enemy left the area recently and quickly, without removing his mines and obstacles. The platoon has all necessary personnel and equipment including, if available, the Mortar Fire-Control System (MFCS); communications with higher, adjacent, and subordinate elements; and guidance on the rules of engagement (ROE). The operational environment might include coalition forces and noncombatants. Some iterations of this task should be performed in limited visibility. Some iterations of this task should be performed in mission-oriented protective posture 4 (MOPP 4).

STANDARDS: The enemy did not surprise either the advance party or the main body. The leader of the advance party (if any) issued a contingency plan to stay-behind unit(s) before departing, and then reconnoitered primary and alternate routes. The advance party or, if no advance party, the mortar platoon, reconnoitered at least two firing positions from which to support the company, and then prepared them. The unit complied with the ROE.

TASK STEPS AND PERFORMANCE MEASURES	GO	NO-GO
*1. Platoon leader gain or maintain situational understanding using information gathered from Force XXI battle command, brigade, and below (FBCB2), the MFCS; FM communications; maps; aerial photos; intelligence summaries; situation reports (SITREPs); current intelligence from the company commander or battalion S2; unmanned aircraft systems (UASs) and other applicable and available information sources. **NOTE:** The MFCS is an automated fire-control system that integrates mortar platoons into current and future fire-support command architecture. The MFCS allows mortar squads to operate semiautonomously. This gives them some flexibility in tactical employment, even though they remain under the control of a central FDC.		
*2. Leader performs a map reconnaissance and plans a ground reconnaissance. (171-610-0001) Takes the following actions:		
a. Chose at least two tentative firing positions.		

TASK STEPS AND PERFORMANCE MEASURES	GO	NO-GO
b. Chose a tentative covered and concealed route(s) to the new firing positions.		
c. Ensures that the reconnaissance plan would allow the unit or advance party to avoid known and suspected contaminated areas and obstacles, when possible.		
d. Identifies the tentative location(s) for pre-positioned supplies and ammunition.		
e. Ensures that the advance party, if used, included—		
(1) A designated leader.		
(2) The ability to compute firing data.		
(3) Communications with the main body.		
(4) A guide for a base mortar squad.		
(5) The ability to detect and monitor chemical and radiological contamination.		
(6) An aiming circle and operator, if the leader wanted a prepared position.		
f. Identifies areas for tentative en route hipshoot firing positions.		
g. Includes, but did not limit the advance party to—		
(1) Personnel.		
(a) Platoon leader.		
(b) Platoon sergeant.		
(c) Section sergeant.		
(2) Equipment.		
(a) A vehicle with a radio, a map, a compass, binoculars, chemical and nuclear detection equipment, and a global positioning system (GPS), if available.		
(b) A map, a grid sheet, overlay paper, a coordinate scale, and a protractor.		
(c) A declinated aiming circle; a field telephone; communications wire; a minefield; and a chemical, biological, radiological, nuclear, and high-yield explosive (CBRNE) warning sign.		
(d) An M23 mortar ballistic computer (MBC), an M16 plotting board, tabular firing tables (TFTs), and an updated weapons-location data card.		
(e) Marking stakes, tape, and a hammer.		
(f) Axes, shovels, an aiming post with lights, and flashlights.		
h. Requests and obtains battalion approval, company approval, or both for the new firing positions and route(s).		

TASK STEPS AND PERFORMANCE MEASURES	GO	NO-GO

3. The mortar unit reconnoiters the route to the firing position. Takes the following actions:
    a. Verifies the cover, concealment, and trafficability of the route.
    b. Selects and reconnoiters an alternate route, if needed.
    c. Reports route changes to the unit's main body and to battalion or company headquarters.
    d. Detects, marks, and reports all chemical or radiological contamination encountered.
    e. If time permits, locates, and marks minefields and obstacles along the route.
    f. Estimates the time needed to reach the new firing position(s), and informs the main body.
    g. Verifies en route hipshoot firing positions.
    h. Reports all key information to the main body before its movement.
    i. Marks routes as needed.
NOTE: Before each mission, the platoon leader designates the duration between position updates. These updates included (at a minimum), the current locations of the platoon leader, squad leaders, and vehicles. Position navigation aids (POSNAV) allow leaders to track their own locations and those of higher, adjacent, and subordinate units; supplement, but do not replace, basic navigational skills; help leaders plan their routes; provide directional information for movement and target acquisition; and augment operational planning graphics such as checkpoints, boundaries, coordination points, and phase lines.
4. The mortar unit reconnoiters and selects the firing position that—
    a. Allows the section or platoon to provide indirect-fire support to the company or battalion, when consistent with the mission.
    b. Satisfies the factors of mission, enemy, terrain, troops, time available, and civil considerations (METT-TC).
    c. Allows the mortars to fire at least one-half of their range to the front of forward-supported elements.		

TASK STEPS AND PERFORMANCE MEASURES	GO	NO-GO
d. Set far enough from forward-supported elements to allow the mortars to place final-protective fires (FPFs) immediately to their front.		
e. Provides maximum coverage of forward-supported elements' frontage, consistent with priority targets and priority of fires.		
f. Allows entry without enemy ground observation.		
g. Covers occupants from direct fire and low-angle indirect fire.		
h. Conceals occupants from air and ground observation.		
i. Avoids high-speed approaches from the forward edge of the battle area (FEBA).		
j. Allows entry and exit by at least two routes.		
k. Permits communications between mortar squads and their fire direction center (FDC).		
l. Offers convenient access to routes for resupply and future displacement.		
m. Masks all mortars, yet left overhead clearance (800 mils to left and right of mounting azimuth). **NOTE:** The MCV has a left and right limit of 4,400 mils total.		
n. Masks all mortars, yet left overhead clearance (800 to 1,511 mils elevation for 60-mm, 81-mm, and 120-mm mortars).		
o. Contain no chemical or radiological contamination.		
5. The advance party prepares the new firing position. (071-074-0017) Takes the following actions:		
a. Establishes local security.		
b. If terrain permits, selects and marks mortar positions at intervals of—		
(1) From 25 to 30 meters (60-mm mortar).		
(2) From 35 to 45 meters (81-mm mortar).		
(3) From 60 meters (120-mm mortar [60 meters distance between positions not required with MFCS]).		
c. Checks firing position and nearby terrain for mines, CBRNE contamination, and enemy forces.		
d. Selects a covered and concealed FDC position.		
e. Prepares the aiming circle.		
f. Marks direction of fire for mortars.		

TASK STEPS AND PERFORMANCE MEASURES	GO	NO-GO
g. Prepares MBC, plotting boards, or both for the new position.		
h. Performs rough lay of the mortar position.		
i. Selects squad sectors, tentative obstacle locations, supplementary defensive positions, and exit routes.		
j. Identifies covered and concealed hide positions.		
k. Identifies alternate and supplementary defensive positions.		
6. The platoon complies with ROE.		
* indicates a leader task step.		

SUPPORTING INDIVIDUAL TASKS

Task Number	Task Title
071-074-0012	Conduct Occupation of a Mortar Firing Position by a Squad
171-610-0001	Perform a Map Reconnaissance
071-326-0513	Select Temporary Fighting Positions

SUPPORTING COLLECTIVE TASKS

Task Number	Task Title
07-2-1342	Conduct Tactical Movement (Platoon-Company)
07-2-5063	Conduct Composite Risk Management (Platoon-Company)
07-3-1351	Occupy a Mortar Firing Position (Section-Platoon)

SUPPORTING BATTLE/CREW DRILLS

Drill Number	Drill Title
07-3-D9501	React to Contact (Visual, IED, Direct Fire [includes RPG])
07-4-D9314	Perform Hasty Lay of a 120-mm Mortar for Hipshoot
07-3-D9504	React to Indirect Fire

TASK: Fire a Mortar Priority Target Mission (Platoon/Squad) (07-3-3099)
(FM 3-22.90)   (FM 3-22.91)

CONDITIONS: The mortar platoon is in a firing position, and has received a request to engage a priority target that is located within the transfer limits of a registration point. The fire-direction center (FDC) has computed and provided the firing data. The platoon has all necessary personnel and equipment including, if available, the Mortar Fire-Control System (MFCS); communications with higher, adjacent, and subordinate elements; and guidance on the rules of engagement (ROE). The operational environment (OE) might include coalition forces or noncombatants. Some iterations of this task should be performed in limited visibility. Some iterations of this task should be performed in mission-oriented protective posture 4 (MOPP 4).

STANDARDS: The mortar platoon initiates accurate fire for effect (FFE) on the priority target within 1 minute (if already laid on it) or 2 minutes (if not already laid on it). The unit engages the enemy or target within the bursting radius for an effective FFE. The unit complies with the ROE.

TASK STEPS AND PERFORMANCE MEASURES	GO	NO-GO
*1. Platoon leader gain or maintain situational understanding using information gathered from Force XXI battle command, brigade, and below (FBCB2), the MFCS; FM communications; maps; aerial photos; intelligence summaries; situation reports (SITREPs); current intelligence from the company commander or battalion S2; and other applicable and available information sources. **NOTE:** The MFCS is an automated fire-control system that integrates mortar platoons into current and future fire-support command architecture. The MFCS allows mortar squads to operate semiautonomously. This gives them some flexibility in tactical employment, even though they remain under the control of a central FDC.		
2. The FDC processes the call for fire.		
a. Determines the initial data and sent it to the guns within 2 minutes of the last element of the call for fire or, if equipped with the Mortar Fire-Control System (MFCS), within 1 minute and 30 seconds. (See the task "Process Call for Fire.") (07-3-5090)		
b. Determines all subsequent corrections within 30 seconds (no MFCS) or 15 seconds (MFCS).		

TASK STEPS AND PERFORMANCE MEASURES	GO	NO-GO
c. Records all firing data on DA Form 2399-R, *Computer's Record.* d. Keeps the DA Form 2188-R, *Data Sheet.* OR DA Form 2188-1-R, *LHMBC/MFCS Data Sheet.* 3. [FDC not present] The unit leader prepares the initial fire command in accordance with steps 2a through 2d. a. Determines the direction and distance from the firing position to the target. b. Authenticates the fire request. c. Determines if the section could accept the fire mission by verifying that— (1) The target was in the unit's area of responsibility. (2) The target did not endanger friendly troops. (3) The fire mission supported the overall mission of the supported unit(s). (4) The required ammunition was available. (5) The target did not conflict with any fire-support coordination measures. (6) Special permission was required and, if necessary, the leader requested permission to fire the mission from the fire-support officer (FSO). **NOTE:** [Section only] The intervals between mortars are 25 to 30 meters for 60-mm mortars, 35 to 45 meters for 81-mm, and 60 meters for 120-mm mortars (60 meters distance between positions not required with MFCS), terrain permitting. 4. The mortar element engages the priority target (section was already laid on the requested priority target). a. Perform Crew Drill, "Load and Fire the Mortar." b. Initiates FFE within 1 minute after receiving the call for fire (no MFCS) or 15 seconds (MFCS). 5. The mortar element engages the priority target (unit not laid on the requested priority target). a. Lay mortar for deflection/azimuth and/or elevation change. (071-086-0003, 071-090-0002) b. Performs Crew Drill, "Load and Fire the Mortar." c. Initiates FFE within 2 minutes after receipt of the call for fire (no MFCS) or one minute and 30 seconds (MFCS).		

TASK STEPS AND PERFORMANCE MEASURES	GO	NO-GO
6. (Live fire only.) The element engages the enemy or point target within the bursting radius. Impact plotters made a target circle by constructing two concentric rings, one with a radius of 50 meters (for the section), and one with a radius of 100 meters (for the platoon). The plotters places the concentric rings wherever the area between them (the actual target "circle") would hold the most FFE rounds. Then, for effective FFE, at least 75 percent of the rounds fired impacted on or between the rings. 7. At the end of the mission, the subordinate element leaders reports the number of rounds expended and the effects of the rounds on enemy personnel or target(s), if known. (171-121-4051) 8. When not firing another fire mission, gunners lay mortars using final protective fire data or the designated priority target data. 9. The platoon complies with the ROE. * indicates a leader task step.		

SUPPORTING INDIVIDUAL TASKS
Task Number   **Task Title**
061-283-6003   Adjust Indirect Fire
301-348-1050   Report Information of Potential Intelligence Value

SUPPORTING COLLECTIVE TASKS
Task Number   **Task Title**
07-3-3072   Fire a Mortar Fire for Effect Mission (Section-Platoon)
07-3-3054   Fire a Mortar Adjust Fire Mission (Section-Platoon)
07-3-5090   Process a Mortar Call for Fire Mission (Section-Platoon)

SUPPORTING BATTLE/CREW DRILLS
Drill Number   **Drill Title**
07-4-D9339   Fire the Mortar
07-4-D9277   Remove a Misfire from 81-mm Mortar

TASK: Conduct Action on Contact (Section-Platoon) (07-3-9013)

(FM 3-21.8)   (FM 3-21.10)

CONDITIONS: The unit conducts operations as part of a higher headquarters (HQ) and receives an operation order (OPORD) or fragmentary order (FRAGO) to conduct a mission at the location and time specified. Unit makes contact with the enemy. Unit receives fires from enemy weapons, visually acquires the enemy, or makes contact with an enemy obstacle. All necessary unit personnel and equipment are available. Indirect fire and close air support (CAS) are available. Unit has established communications with required headquarters/units at all echelons. Unit has received guidance on the rules of engagement (ROE). Coalition forces and noncombatants may be present in the operational environment. Civilians, government agencies, nongovernment organizations (NGOs); and local and international media may be in the area. Some iterations of this task should be performed under the mission variables of: mission given, enemy forces and their capabilities, terrain and weather effects, troops available, time available to execute the operation, and civil considerations (METT-TC), which are conditions that aid or limit performance. Some iterations of this task should be performed in mission-oriented protective posture 4 (MOPP 4).

STANDARDS: The unit conducts action on contact according to the standing operating procedures (SOPs), the order, higher commander's guidance, and/or the tactical situation. The element in contact deploys and reports, initiates fires to destroy or suppress the enemy, and chooses a course of action (COA). The unit conducts necessary battles drills, maneuvers to bypass, withdraw, or assault the enemy position. The unit complies with the ROE, the five steps of actions on contact, mission instructions, higher HQ order, and other special orders. The unit treats local inhabitants with respect.

TASK STEPS AND PERFORMANCE MEASURES	GO	NO-GO
PLAN 1. Planning for actions on contact begin when the unit receives mission orders. The unit conducts troop-leading procedures (TLPs) and plan for all possible contact it expects to encounter. PREPARE		

TASK STEPS AND PERFORMANCE MEASURES	GO	NO-GO
*2. Unit leaders complete their plans and issue the order. The unit initiates rehearsals of tactical movement and action drills based on likely enemy contact. They should consider the following: a. Rehearsals should be matched to the actual terrain and anticipated actions on contact with the enemy b. Rehearsals allow the unit to begin preparing for the mission, and must include the five following steps of action on contact: (1) Deploy and report. When the unit encounters an enemy unit or obstacle, it deploys to a covered position that provides observation and fields of fire. (2) Evaluate and develop the situation. While the unit deploys, the unit leaders evaluate the situation and continue to develop it. (3) Choose a COA. After developing the situation and determining that enough information was gathered to make a decision, the unit leaders select a COA that meets both the requirements of the commander's intent and is within the unit's capabilities. (4) Execute the selected COA. When executing the selected COA, the unit transitions from movement to maneuver. (5) Recommend a COA to the higher HQ. Once the unit selects a COA, keeping in mind the commander's intent. It informs the commander, who has the option of disapproving it based on its impact on the overall mission. EXECUTE 3. The element in contact deploys and reports by taking the following actions: a. Reacting when contact entails direct fire, which can include the following actions: (1) Return fire immediately to destroy or suppress the enemy. (2) Deploy to available covered and concealed positions. (3) If the direct fire is identified or suspected to be sniper fire, the unit should: (a) Take cover immediately.		

TASK STEPS AND PERFORMANCE MEASURES	GO	NO-GO
(b) Employ smoke to obscure sniper's view and conceal unit's movement.		
(c) Identify sniper location. Look for reflections, dust clouds, or muzzles flash using all available resources (binoculars, thermal imager, and weapons scopes).		
(d) Return well-aimed fire according to ROE to destroy or suppress the sniper.		
(4) Close hatches and activate on-board self-protection measures as appropriate, if applicable.		
(5) Call for indirect fire, CAS, or attack aviation support, if available.		
(6) Conduct battle drills, as needed.		
(7) Maintain visual contact with the enemy while continuing to develop the situation through reconnaissance or surveillance.		
(8) Maintain cross talk with all unit elements.		
b. Reacting to visual contact (Element is NOT in immediate danger), which can include the following actions:		
(2) Maintain visual contact.		
(3) Maintain cross talk with all unit elements.		
(4) Conduct further actions as directs by unit leaders, which might include bypass of enemy position		
c. Reacting to visual contact (Element is in immediate danger.), which can include the following actions:		
(1) Initiate fires to destroy or suppress the enemy.		
(2) Deploy to covered and concealed positions.		
(3) Close hatches, if applicable.		
(4) Activate on-board self-protection measures, as appropriate.		
(5) Maintain cross talk with all unit elements.		
(6) Conduct further actions, as directed by unit leaders.		
d. Reacting when contact was indirect fires (observed or receiving), which can include the following actions:		
(1) Use evasive actions to avoid impact area.		
(2) Move quickly to clear impact area.		

TASK STEPS AND PERFORMANCE MEASURES	GO	NO-GO
(3) Close hatches, if applicable.		
(4) React to chemical and or biological attack, if necessary, and immediately conducts chemical, biological, radiological, nuclear, and high-yield explosive (CBRNE) reconnaissance, as required.		
(5) Maintain cross talk with all unit elements.		
(6) Conduct further actions, as directed, by unit leaders.		
e. Reacting when contact was with an obstacle, which can include the following actions:		
(1) Deploy to covered and concealed positions.		
(2) Maintain cross talk with all unit elements.		
(3) Call for immediate smoke on the far side of the obstacle to conceal deployment of the unit, if required.		
(4) Make a recommendation to higher headquarters (bypass or breach).		
(a) Bypass, if possible.		
(b) Breach, if required.		
f. Reacting to visual contact of enemy or unknown aircraft (Element is in immediate danger.), which can include the following actions:		
(1) Initiate fires to destroy or cause aircraft to depart the area.		
(2) Deploy to covered and concealed positions.		
(3) Close hatches, if applicable.		
(4) Activate on-board, self-protection measures, as appropriate.		
(5) Maintain cross talk with all unit elements.		
(6) Conduct further actions, as directed, by unit leaders.		
g. Reacting to visual contact of enemy or unknown aircraft (Element is not in immediate danger.), which can include the following actions:		
(1) Deploy to covered and concealed positions.		
(2) Maintain visual contact.		
(3) Maintain cross talk with all unit elements.		
(4) Conduct further actions, as directed, by unit leaders.		
*4. Unit leaders evaluate the situation by taking the following actions:		

TASK STEPS AND PERFORMANCE MEASURES	GO	NO-GO
a. Confirming friendly and enemy situations by taking the following actions:		
(1) Receive an updated report showing the location of forward and adjacent friendly elements, if applicable.		
(2) Receive an updated enemy situational template for added fratricide prevention to support risk management initiatives and increased force protection, if applicable.		
b. Conducting reconnaissance to fully develop the situation.		
c. Determining enemy size, composition, activity, orientation, and location of weapon systems.		
d. Searching for antitank ditches, minefields, wire, or other obstacles that could define an engagement area.		
e. Searching for the flanks of the enemy and any elements that could mutually support enemy positions.		
f. Calling for indirect fire, CAS, or attack aviation support, if available.		
g. Analyzing element spot reports (SPOTREPs) and other tactical information, as required, making an assessment of the situation.		
h. Sending updated SPOTREPs to higher HQ based on a fully developed situation.		
i. Directing the actions of elements not in contact in a manner that supports the elements in contact or to continue the mission according to the OPORD and/or FRAGO.		
*5. Unit leaders disseminate reports (if applicable), overlays, and other pertinent information to each element to keep them abreast of the situation.		
*6. Unit leaders select an appropriate COA based on the commander's intent; METT-C; analysis of the situation; and input from elements in contact. COAs should follow one of the proceeding actions:		
a. Direct the unit to execute the original COA (as previously addressed in the OPORD) if it is consistent with the commander's intent/concept and is within the unit's capability. OR		

TASK STEPS AND PERFORMANCE MEASURES	GO	NO-GO
b. Issue a FRAGO to refine the plan based on the situation, ensuring it supports the commander's intent. OR		
c. Direct the unit to execute tactical movement (employing bounding overwatch and support by fire within the unit) and reconnaissance by fire to further develop the situation. OR		
d. Direct the unit to establish a hasty defense/support by fire position and take further guidance from commander. OR		
e. Choose an alternative COA based on evaluation and development of the situation.		
*7. Unit leaders recommend alternative COAs (if situation dictates a change to the original plan), and take the following actions:		
a. Send recommendation to the higher HQ commander.		
b. Receive orders to execute the COA selected by the higher HQ commander.		
c. Use cross talk with other units, as necessary, to obtain support (unit leaders or designates representatives).		
*8. Unit leaders direct the unit to execute the chosen COA based on the situation or leader's order. The COA should follow one of the proceeding actions:		
a. Direct the unit to destroy an inferior force. OR		
b. Direct unit to conduct overwatch and or support by fire. OR		
c. Direct unit to conduct an attack by fire. OR		
d. Direct unit to assault an enemy position. OR		
e. Direct unit to break contact and conduct bypass operations. OR		
f. Direct unit to conduct reconnaissance by fire. OR		

TASK STEPS AND PERFORMANCE MEASURES	GO	NO-GO

g. Direct unit to conduct defense of a battle position. OR
   h. Direct unit to breach an obstacle.
ASSESS
*9. Unit leaders, or designated representatives, keep the higher command informed throughout the operation by taking the following actions:
   a. Sending updated situation reports (SITREPs) and or SPOTREPs, as needed.
   b. Reporting completion of the operation.
10. The unit conducts consolidation and reorganization, as necessary, to continue operations.
11. The unit complies with ROE.
*indicates a leader task step.		

SUPPORTING INDIVIDUAL TASKS

Task Number	Task Title
061-283-6003	Adjust Indirect Fire
301-348-1050	Report Information of Potential Intelligence Value
113-571-1022	Perform Voice Communications
171-121-3009	Control Techniques of Movement
071-420-0005	Conduct the Maneuver of a Platoon
171-620-0094	Conduct Consolidation and Reorganization Activities at Company-Troop Level
071-326-5502	Issue a Fragmentary Order

SUPPORTING COLLECTIVE TASKS

Task Number	Task Title
07-2-1256	Conduct an Attack by Fire (Platoon-Company)
07-2-1342	Conduct Tactical Movement (Platoon-Company)
07-2-1477	Breach an Obstacle (Platoon-Company)
07-2-3000	Conduct Support by Fire (Platoon-Company)
07-2-9002	Conduct a Bypass (Platoon-Company)
07-2-9003	Conduct a Defense (Platoon-Company)
08-2-0003	Treat Casualties
08-2-0004	Evacuate Casualties
19-3-3107	Process Detainee(s) at Point of Capture (POC)

SUPPORTING BATTLE/CREW DRILLS

Drill Number   **Drill Title**

07-3-D9501   React to Contact (Visual, IED, Direct Fire [includes RPG])

17-3-D8004   React to an Air Attack Drill

TASK: Fire a Mortar Adjust Fire Mission (Platoon/Squad) (07-3-3054)

(FM 3-22.90)   (FM 3-22.91)

CONDITIONS: The mortar platoon's fire-direction center (FDC) is in a firing position and has received a request to adjust fire. The platoon has all necessary personnel and equipment including, if available, the Mortar Fire-Control System (MFCS); communications with higher, adjacent, and subordinate elements; and guidance on the rules of engagement (ROE). The operational environment (OE) might include coalition forces or noncombatants. Some iterations of this task should be performed in limited visibility. Some iterations of this task should be performed in mission-oriented protective posture 4 (MOPP 4).

STANDARDS: The mortar platoon initiates an accurate fire for effect (FFE) within 7 minutes after receiving fire request or, if equipped with the MFCS, within 4 minutes. The platoon engages the enemy or target within the bursting radius and complies with the ROE.

TASK STEPS AND PERFORMANCE MEASURES	GO	NO-GO
*1. Platoon leader gains or maintains situational understanding using information gathered from Force XXI battle command, brigade, and below (FBCB2), MFCS; FM communications; maps; aerial photos; intelligence summaries; situation reports (SITREPs); current intelligence from the company commander or battalion S2; unmanned aircraft systems (UAS) and other applicable and available information sources.		
NOTE: The MFCS is an automated fire-control system that integrates mortar platoons into current and future fire-support command architecture. The MFCS allows mortar squads to operate semiautonomously. This gives them some flexibility in tactical employment, even though they remain under the control of a central FDC.		
2. The FDC processes the adjust-fire request. (07-3-5090)		
a. Determines the initial data and sent it to the guns within 2 minutes of the last element of the call for fire or, if equipped with MFCS, within 1 minute and 30 seconds.		
b. Determines all subsequent corrections within 30 seconds (no MFCS) or 15 seconds (MFCS).		
c. Records all firing data on DA Form 2399-R, *Computer's Record.*		

TASK STEPS AND PERFORMANCE MEASURES	GO	NO-GO

d. Keeps the current DA Form 2188-R, *Data Sheet.*
OR
   DA Form 2188-1-R, *LHMBC/MFCS Data Sheet.*
3. [FDC not present] The unit leader prepares the initial fire command according to steps 2a through 2d.
   a. Determines the direction and distance from the firing position to the target.
   b. Authenticates the fire request.
   c. Determines if the section can accept the fire mission by verifying that—
      (1) The target was located in the element's area of responsibility.
      (2) The target did not endanger friendly troops.
      (3) The fire mission supported the overall mission of the supported unit(s).
      (4) The required ammunition was available.
      (5) The target did not conflict with any fire-support coordination measures.
      (6) Leader cleared the fire-support request (FR) according to the unit tactical standing operation procedures (TSOP).
NOTE: The intervals between mortars are 25 to 30 meters for 60-mm mortars, 35 to 45 meters for 81-mm, and 60 meters for 120-mm mortars (60 meters distance between positions not required with MFCS), terrain permitting.
4. The mortar element adjusted fire.
   a. Lay mortar for deflection/azimuth and/or elevation change.
   b. [Base element] Performs Crew Drill, "Load and Fire the Mortar."
   c. [Base element] Fires initial adjusting round within 2 minutes after receipt of the target location or, if equipped with MFCS, within one minute and 30 seconds. (07-3-3054, 07-3-3072)
   d. [Base element] Fires subsequent adjusting rounds within 1 minute after impact of the previous round or, if equipped with MFCS, within 15 seconds.
5. The mortar element conducts FFE.
   a. Performed Crew Drills, "Load and Fire the Mortar."		

TASK STEPS AND PERFORMANCE MEASURES	GO	NO-GO
b. Initiates FFE within 5 minutes after receipt of the target location or, if equipped with MFCS, within 3 minutes. **NOTE:** The element engages the enemy or target within the bursting radius. 6. At the end of the mission, subordinate element leaders reported the number of rounds expended and the effects of the rounds on enemy personnel or target(s). (171-121-4051) 7. When not firing another mission, gunners laid the mortars using final protective fire data or the designated priority target data. 8. The platoon complied with the ROE. * indicates a leader task step.		

SUPPORTING INDIVIDUAL TASKS

Task Number	Task Title
061-283-6003	Adjust Indirect Fire
301-348-1050	Report Information of Potential Intelligence Value

SUPPORTING COLLECTIVE TASKS

Task Number	Task Title
07-3-3054	Fire a Mortar Adjust Fire Mission (Section-Platoon)
07-3-3072	Fire a Mortar Fire for Effect Mission (Section-Platoon)
07-3-5090	Process a Mortar Call for Fire Mission (Section-Platoon)

SUPPORTING BATTLE/CREW DRILLS

Drill Number	Drill Title
07-4-D9339	Fire the Mortar
07-4-D9275	Lay 81mm Mortar for Large Deflection and Elevation Change

TASK: Fire a Mortar Fire for Effect Mission (Platoon/Squad) (07-3-3072)
(FM 3-22.90)   (FM 3-22.91)

CONDITIONS: The mortar platoon is in a firing position and has received a request to fire for effect, without adjusting on a target within the transfer limits of the registration point. The platoon has all necessary personnel and equipment including, if available, the Mortar Fire-Control System (MFCS); communications with higher, adjacent, and subordinate elements; and guidance on the rules of engagement (ROE). The operational environment (OE) might include coalition forces or noncombatants. Some iterations of this task should be performed in limited visibility. Some iterations of this task should be performed in mission-oriented protective posture 4 (MOPP 4).

STANDARDS: The mortar platoon initiates an accurate FFE within 4 minutes of fire request or, if equipped with the MFCS, within three minutes and 30 seconds. The unit engages the enemy or target within the bursting radius. The platoon complies with the ROE.

TASK STEPS AND PERFORMANCE MEASURES	GO	NO-GO
*1. Platoon leader gain or maintain situational understanding using information gathered from Force XXI battle command, brigade, and below (FBCB2), MFCS; FM communications; maps; aerial photos; intelligence summaries; situation reports (SITREPs); current intelligence from the company commander or battalion S2; unmanned aircraft systems (UASs) and other applicable and available information sources.		
NOTE: The MFCS is an automated fire-control system that integrates mortar platoons into current and future fire-support command architecture. The MFCS allows mortar squads to operate semiautonomously. This gives them some flexibility in tactical employment, even though they remain under the control of a central FDC.		
2. The fire direction center (FDC) processes the fire for effect (FFE) request.		
a. Determines the initial data and sent it to the guns within 2 minutes of the last element of the call for fire or, if equipped with MFCS, within 1 minute and 30 seconds. (07-3-5090)		
b. Determines all subsequent corrections within 30 seconds (no fire control system [FCS]) or 15 seconds (MFCS).		

TASK STEPS AND PERFORMANCE MEASURES	GO	NO-GO
c. Records all firing data on DA Form 2399-R, *Computer's Record*. d. Keeps the current DA Form 2188-R, *Data Sheet*. OR DA Form 2188-1-R, *LHMBC/MFCS Data Sheet*. 3. (FDC not present.) The unit leader prepares the initial fire command in accordance with steps 2a through 2d. a. Determines the direction and distance from the firing position to the target. b. Clears the fire-support request according to the unit tactical standing operation procedures (TSOP). 4. The mortar unit conducts FFE. a. Lay mortar for deflection/azimuth and/or elevation change. (071-090-0002, 071-323-4102) b. Performs Crew Drill, "Load and Fire the Mortar." c. Initiates FFE within 4 minutes after receipt of the target location or, if equipped with MFCS, within 3 minutes. 5. [Live fire only] The element engages the enemy or point target within the bursting radius. Impact plotters made a target circle by constructing two concentric rings, one with a radius of 50 meters (for the section), and one with a radius of 100 meters (for the platoon). The plotters places the concentric rings wherever the area between them (the actual target "circle") would hold the most FFE rounds. Then, for effective FFE, at least 75 percent of the rounds fires impacted on or between the rings. **NOTE:** For example, if 3 rounds are fired, then 3 times 0.75 (75 percent) equals 2.5, which rounds down to 2. Therefore, 2 out of 3 rounds fired must impact within the target circle (the area between the two rings). 6. At the end of the mission, the subordinate element leaders reports the number of rounds expended and the effects of the rounds on enemy personnel or target(s). (171-121-4051) 7. When not firing another mission, gunners lay the mortars using final protective fire data or designated priority target data. 8. The platoon complies with the ROE. *indicates a leader task step.		

SUPPORTING INDIVIDUAL TASKS

Task Number   **Task Title**
061-283-6003   Adjust Indirect Fire
301-348-1050   Report Information of Potential Intelligence Value

SUPPORTING COLLECTIVE TASKS

Task Number   **Task Title**
07-3-5090   Process a Mortar Call for Fire Mission (Section-Platoon)
07-3-3054   Fire a Mortar Adjust Fire Mission (Section-Platoon)
07-3-3072   Fire a Mortar Fire for Effect Mission (Section-Platoon)

SUPPORTING BATTLE/CREW DRILLS

Drill Number   **Drill Title**
07-4-D9339   Fire the Mortar

TASK: Conduct Logistics Package (LOGPAC) Support (63-2-4546)

(ADP 4-0)   (FM 5-19)   (ATP 4-11)

CONDITIONS: Unit receives an operations order (OPORD) and/or fragmentary order (FRAGO) to conduct resupply operations upon the arrival of the logistics package (LOGPAC), or the commander determines that routine or emergency resupply is necessary. The unit has established communications with subordinate, adjacent and higher headquarters (HQ), and is passing information according to the tactical standing operating procedure (TSOP). Unit has been provided guidance on the rules of engagement (ROE). Coalition forces and noncombatants may be present in the operational environment. This task is performed under all day and night environmental conditions. Threat capabilities cover a full spectrum to include: information gathering; hostile force sympathizers; terrorist activities to include: suicide bombings; and conventional, air supported, and reinforced squad operations in a chemical, biological, radiological, nuclear, and high-yield explosive (CBRNE) environment. Some iterations of this task should be performed in mission-oriented protective procedure 4 (MOPP 4).

STANDARDS: Unit requests supplies/services necessary to restore it to fully mission capable (FMC) status. Receives supplies and services as available and conducts distribution as needed to subordinate elements. Unit completes resupply operations within the time specified in the OPORD and/or FRAGO, or command guidance. Unit complies with ROE. No friendly unit suffers casualties or equipment damage as a result of fratricide.

TASK STEPS AND PERFORMANCE MEASURES	GO	NO-GO

*1. The executive officer (XO)/first sergeant (1SG) monitors supply status and reports status as required by unit tactical standing operating procedure (TSOP). (101-92A-4216)
    a. Compile accurate supply status (by class) from leaders of each platoon/section/element. Reports cover the following supply classes:
        (1) Class I (Rations).
        (2) Class II (Supplies and Equipment).
        (3) Class III (Petroleum, Oil, and Lubricants [POL] products).
        (4) Class IV (Construction/Barrier Materials).
        (5) Class V (Ammunition).
        (6) Class VI (Personnel Demand Items).
        (7) Class VII (Major End Items).		

TASK STEPS AND PERFORMANCE MEASURES	GO	NO-GO
(8) Class VIII (Medical Supplies).		
(9) Class IX (Repair Parts).		
(10)Class X (Nonmilitary Program Materials such as agriculture and economic development).		
(11)Water.		
b. Submit consolidated logistical status (LOGSTAT) report through unit commander to higher HQ S-4 and/or forward support company (FSC)		
2. Unit reports personnel status to the higher HQ S-1 using personnel status (PERSTAT) report, requests replacements, and processes reassignment/ replacements.		
a. Platoon sergeants (PSGs) report personnel strength/losses (with battle roster numbers) to platoon/element leader and XO/1SG using PERSTAT.		
b. 1SG compiles report of personnel strength, losses, and battle roster changes and submits roll-up PERSTAT through the company commander to the higher HQ S-1.		
c. 1SG and PSGs reassign remaining personnel to ensure key positions are filled and critical weapons are manned.		
d. 1SG and PSGs assign replacements using the same criteria.		
e. Notifies s operations officer (SOO) when LOGPAC Operations vehicles are fully loaded and ready to move.		
f. Verifies that trail party is equipped to recover vehicles that develop maintenance problems during the combat resupply operations convoy.		
3. Unit reports vehicle status and requests resupply or other support as needed.		
a. PSGs and section leaders report vehicle and equipment status to include battle damage assessment (BDA), to platoon leaders and XO/1SG.		
b. PSGs and section leaders report maintenance, recovery, and evacuation support requirements to platoon leaders and XO/1SG.		
c. XO/1SG compiles platoon/section reports/requests and maintenance forecast and submits them to the higher HQ S-4 and/or supporting maintenance unit.		

TASK STEPS AND PERFORMANCE MEASURES	GO	NO-GO
d. They forward SP crossing report to HQ when unit elements have crossed the SP using FBCB2, MTS, or FM radio. e. They employ correct signal operating instructions/signal supplemental instructions (SOI/SSI) codes in all transmissions. f. They enforce march discipline using FBCB2, MTS, FM radio, or proper visual signals. *4. XO/1SG coordinate logistical package (LOGPAC) with higher HQ S-4 and/or forward support company (FSC) (191-379-4407). He takes the following actions: a. Verify status of resupply/support requests b. Coordinate actions at the logistics release point (LRP). c. Assume position(s) along march route that provides command presence at points of decision for reaction to changing tactical situation. d. Maintain situational awareness at all times using FBCB2 and MTS. e. Forward enroute CBRN information. f. Enforce all movement policies defined in the TSOP and movement order, with emphasis on formation, distances, speeds, passing procedures, and halts. g. Report all threat sightings using SALUTE (size activity location unit time equipment) Report format. h. Adjust formation distances and speed consistent with CBRN, terrain, and light conditions. i. Enforce security measures, with emphasis on air guard's surveillance, manning of automatic weapons, and concealment of critical cargo. j. Inform vehicle operators by FBCB2, radio, MTS, or proper visual signals, any violations of march discipline, security procedures, or changes to established orders. k. Enforce communications security (COMSEC) measures to include radio silence periods according to the OPORD and SOI/SSI. 5. The supply sergeant (under the supervision of the Headquarters and Headquarters Company [HHC] or FSC commander) assembles the LOGPAC. He takes the following actions:		

TASK STEPS AND PERFORMANCE MEASURES	GO	NO-GO
a. Obtain requested supplies from FSC or higher HQ S-4.		
b. Obtain Class II, IV, VI, and VII supplies from higher HQ S-4 personnel.		
c. Consolidate replacement personnel and those returning from medical treatment.		
d. Consolidate vehicles returning from maintenance.		
e. Obtain mail from higher HQ S-1.		
f. Obtain personnel action documents from S1 section (to include award, finance, and legal documents).		
*6. 1SG/XO meets LOGPAC elements at the LRP. He takes the following actions:		
a. Move to the LRP and meets the supply sergeant and LOGPAC.		
b. Supervise actions at LRP as coordinated and/or specified by unit SOP.		
c. Occupy hasty defensive positions with 360-degree protective coverage (passengers).		
d. Report scheduled halt to HQ.		
e. Direct performance of preventive maintenance checks and services (PMCS) on vehicles.		
f. Inspect vehicle loads for safety and security.		
g. Begin departure at time specified by orders or designated by platoon leader.		
h. Report resumption of march to headquarters.		
*7. 1SG/XO coordinates unit resupply. He takes the following actions:		
a. Determine method of resupply (service station or tailgate).		
b. Determine location(s) of resupply		
c. Determine unit priority for resupply if all required supplies/services are not available.		
d. Determine unit order of resupply to include attachments.		
e. Execute LOGPAC operations according to TSOP or issues FRAGO notifying unit of changes to normal LOGPAC operations.		
f. Reports resumption of march to higher HQ.		
8. The unit receives service station resupply if applicable. The following actions are taken:		

TASK STEPS AND PERFORMANCE MEASURES	GO	NO-GO
a. 1SG/XO escort LOGPAC move to designated resupply location along covered and concealed route.		
b. The unit security element conducts link-up with 1SG/XO and LOGPAC to organize resupply site, establishing security and use available cover and concealment.		
c. 1SG/XO issues FRAGO to PSGs and section sergeants on the organization of the resupply site, specific locations of medics, maintenance, supply points, mortuary affairs collection points and enemy prisoners of war (EPW) collection points.		
d. Support platoons/sections/elements conduct tactical movement to resupply site.		
e. Support platoons/sections/elements conduct appropriate actions of service station resupply as directed by the commander and/or unit SOP.		
*9. Convoy commander conducts night convoy. He takes the following actions:		
a. Brief drivers on night conditions.		
b. Provide visual adjustment period if march began during daylight.		
c. Prepare vehicles for blackout conditions according to the TSOP.		
d. Maintain prescribed interval between vehicles.		
e. Direct the wearing of night vision goggles (selected personnel).		
f. Direct the wearing of regular eye protection goggles (all other personnel).		
g. Enforce the use of ground guides during poor visibility periods.		
*10. Convoy commander conducts convoy through an urban area. He takes the following actions:		
a. Verify all weight, height, and width restrictions along route of march.		
b. Employ close column formation.		
c. Ensure that vehicle drivers obey traffic control directions unless escorted by military or host nation (HN) police.		
d. Employ directional guide's at all critical intersections.		

TASK STEPS AND PERFORMANCE MEASURES	GO	NO-GO

*11.   The convoy commander coordinates/monitors actions at the designated LRP. He takes the following actions:
    a.   Verify that lead vehicle has arrived at the LRP.
    b.   Verify that all vehicles have arrived at the LRP.
    c.   Release unit serials to the supported unit's 1SG or his/her designated represented representative.
    d.   Direct unit serial reassembly at the LRP following unit resupply actions.
    e.   Lead reassembled combat resupply operations convoy back to release point (RP) in the battalion field trains area.
    f.   Ensure that all back haul logistics commodities arrive at the proper location.
    g.   Forward situation report (SITREP) to headquarters using FBCB2, MTS, or radio.
* indicates a leader task step.		

SUPPORTING INDIVIDUAL TASKS

Task Number	Task Title
101-92A-4216	Coordinate Logistical Requirements
191-379-4407	Plan Convoy Security Operations
101-92A-8030	Manage Unit Supply Operations

SUPPORTING COLLECTIVE TASKS

Task Number	Task Title
63-2-4519	Transport Supplies, Equipment, and Personnel
63-2-4000	Coordinate Replenishment/Sustainment Operations
07-2-5036	Conduct Coordination (Platoon-Company)

SUPPORTING BATTLE/CREW DRILLS

Drill Number	Drill Title
07-3-D9501	React to Contact (Visual, IED, Direct Fire [includes RPG])

TASK: Conduct Operational Decontamination (03-2-9224)

(FM 3-11.5)   (FM 3-11)

CONDITIONS: The element is operating in a contaminated environment. Performance degradation from mission-oriented protective posture 4 (MOPP 4) is increasing and protective gear is in danger of contamination. The time and tactical situation permit the element to conduct operational decontamination. Replacement protective gear is available for each Soldier. For a nonsupported decontamination, decontamination equipment and supplies are available and operational. For a supported decontamination, an operational decontamination unit is available and is tasked to provide decontamination support. This task is always performed in MOPP 4.

STANDARDS: The element decontaminates individual gear and conducts MOPP 4 gear exchange (using the buddy team, triple team, or individual (emergency) method) without sustaining additional casualties from chemical, biological, radiological, and nuclear (CBRN) contamination. The element limits the contamination transfer hazard by removing gross chemical contamination from equipment. The element reduces radiological contamination to negligible risk levels according to the element's tactical standing operating procedure (TSOP) and field manual (FM) guidance and/or reduces chemical and biological (CB) contamination to accelerate the weathering process and eventually provide temporary relief from MOPP 4.

TASK STEPS AND PERFORMANCE MEASURES	GO	NO-GO

*1. The element leader determines the extent of the contamination and establishes the priorities for decontamination. He takes the following actions:
  a.   Receives input from subordinate leaders and staff.
  b.   Directs decontamination priorities.
  2. The element submits a request for decontamination to higher headquarters (HQ). The request should include, as a minimum, the following:
NOTE: Decontamination operations should be done between one and six hours after becoming contaminated.
  a.   The designation of the contaminated element.
  b.   The location of the contaminated element.
  c.   The frequency and call sign of the contaminated element.
  d. The time that the element became contaminated.		

TASK STEPS AND PERFORMANCE MEASURES	GO	NO-GO

e.   The number of personnel requiring a MOPP gear exchange.
   f.   The number of vehicles and equipment (by type) that are contaminated.
   g.   The type of contamination.
   h.   Special requirements (such as a patient decontamination station, recovery assets, and an element decontamination team).
  3. The element coordinates with higher HQ. It takes the following actions:
   a.   Obtains permission to conduct decontamination.
   b.   Obtains the necessary support to conduct decontamination.
   c.   Selects the link up point to meet supporting units (a company supply section, a company or battalion power-driven decontamination equipment [PDDE] crew, or a decontamination squad or platoon).
   d.   Coordinates with supporting elements.
   e.   Requests replacement MOPP gear.
   f.   Coordinates with supporting units to determine if they need to exchange MOPP gear also.
*4. The element leader and CBRN specialists select a site to conduct the operation and ensure that the selected site provides:
   a.   Adequate overhead concealment.
   b.   Good drainage.
   c.   Easy access and exit routes (off the main routes).
   d.   Close proximity to a water source large enough to support vehicle wash-down (plan for 100 gallons per vehicle).
   e.   A large enough area to accommodate the elements involved in operational decontamination (110 square meters for both the vehicle wash-down site and the MOPP gear exchange site).
  5. The element coordinates for operational decontamination support (a company or battalion PDDE crew or a decontamination unit). It takes the following actions:		

TASK STEPS AND PERFORMANCE MEASURES	GO	NO-GO

a.   Notifies higher HQ of the site selected for the operational decontamination.
     b.   Establishes communications with the decontamination unit.
     c.   Ensures that the decontamination unit knows the link up locations and the selected decontamination site.
 6. The element and supporting units move to the decontamination site. They take the following actions:
     a.   Meet at the link up point as coordinated.
     b.   Provide security at the link up point and the decontamination site.
 7. The element prepares for operational decontamination. It takes the following actions:
     a.   Sets up the decontamination site.
       (1) The supporting decontamination unit crew sets up a vehicle wash-down site.
       (2) The contaminated element sets up a MOPP gear exchange site no less than 50 meters upwind from the vehicle wash-down at a 45 degree angle.
       (3) The remainder of the element prepares its equipment for decontamination.
     b.   Conducts preparatory actions in the predecontamination marshalling area.
       (1) Vehicle crews (except operators) dismount unless they have an operational overpressure system and an uncontaminated interior.
       (2) Dismounted crews remove mud and camouflage from vehicles.
 NOTE: The contaminated element provides personnel to do this when crews do not dismount.
       (3) Separated vehicles and dismounted crews:
         (a) Ensure that vehicle operators are briefed (include the use of overhead cover and concealment and proper intervals).
         (b) Ensure that vehicles are buttoned up (all doors, hatches, and other openings closed or covered to include muzzles).
       (4) Moves vehicles (with operators) to the vehicle wash-down site.		

TASK STEPS AND PERFORMANCE MEASURES	GO	NO-GO
(5) Moves dismounted crews and all other Soldiers in the contaminated element to the MOPP gear exchange site.		
*8. The noncommissioned officer in charge (NCOIC) of the decontamination unit supervises the operation of the vehicle wash-down site. He ensures that:		
a. Vehicle operators maintain proper intervals between vehicles while processing through the wash-down station.		
b. Decontamination crew washes vehicles properly.		
(1) Starts at the top and work down.		
(2) Sprays hot, soapy water for 2 to 4 minutes per vehicle.		
(3) Wears a toxicological agent-protective (TAP) or wet-weather gear over MOPP gear.		
(4) Monitors water consumption.		
c. Operators move to the MOPP gear exchange after vehicle has been washed down.		
d. Operators move to the assembly area.		
9. The contaminated element conducts MOPP gear exchange. It takes the following actions:		
a. Prepares the equipment decontamination station with super tropical bleach (STB) dry mix.		
b. Briefs MOPP gear exchange participants on the procedures to be followed.		
c. Places the decontaminated individual equipment on a clean surface (such as plastic, a poncho, or similar material).		
d. Exchanges MOPP gear using the buddy team, triple team or individual (emergency) method.		
NOTE: The individual emergency method is used only when a person does not have a buddy to help and the risks of MOPP failure demands that an MOPP exchange occur.		
e. Moves to the assembly area after they complete the MOPP gear exchange.		
10. Supporting elements process through the MOPP gear exchange site.		

TASK STEPS AND PERFORMANCE MEASURES	GO	NO-GO
11. The supporting decontamination element cleans and marks the site and reports the area of contamination using a nuclear, biological, chemical CBRN 5 report to higher HQ.		
*12. The element leader accounts for all personnel and equipment after completing the operational decontamination.		
*13. The element leader reports to higher HQ. He takes the following actions:		
a. Reports the completion of decontamination and the location of the vehicle wash-down and MOPP gear exchange decontamination sites.		
b. Requests permission to perform unmasking procedures if no hazards are detected through testing.		
c. Determines the adequacy of the decontamination and adjusts the MOPP level as required (after obtaining approval from higher HQ).		
14. The element continues its mission.		
* indicates a leader task step.		

SUPPORTING INDIVIDUAL TASKS

Task Number	Task Title
031-503-1019	React to Chemical or Biological (CB) Hazard/Attack
031-503-1021	Mark NBC Contaminated Area
031-503-1031	Use the Chemical Agent Monitor
031-503-1035	Protect Yourself from Chemical and Biological (CB) Contamination Using Your Assigned Protective Mask
031-503-1037	Detect Chemical Agents Using M8 or M9 Detector Paper
031-507-3014	Supervise Decontamination Procedures
113-571-1022	Perform Voice Communications
113-573-8006	Use an Automated Signal Operation Instruction (SOI)
551-721-1352	Perform Preventive Maintenance Checks

SUPPORTING COLLECTIVE TASKS

Task Number	Task Title
07-2-5009	Conduct a Rehearsal (Platoon-Company)
07-2-5063	Conduct Composite Risk Management (Platoon-Company)

07-2-5081   Conduct Troop-Leading Procedures (Platoon-Company)
07-2-6063   Maintain Operations Security (Platoon-Company)

SUPPORTING BATTLE/CREW DRILLS
Drill Number   Drill Title
07-3-D9483   React to Nuclear Attack
03-3-D0035   React to a Chemical Attack

TASK: Treat Casualties (08-2-0003)

(FM 4-25.11)   (AR 190-8)   (FM 4-02.7)

CONDITIONS: The unit has sustained casualties. The unit has medical treatment personnel and/or combat lifesavers. Threat force contact has been broken. Soldiers are wounded and may have chemical contamination or non-battle injuries. Unit personnel perform first aid (self-aid/buddy aid) treatment. The unit has analog and/or digital communications. A higher headquarters (HQ) operation order (OPORD) is available. Unit and higher HQ standing operating procedures (SOPs) are available. A treatment plan is available. This task is performed under all environmental conditions. The unit may be subject to attack by threat forces, including air; ground; chemical, biological, radiological, and nuclear (CBRN); or directed energy (DE) attack. Simplified collective protective equipment (SCPE) is on hand and/or field-expedient and natural shelters are available. Some iterations of this task should be performed in mission-oriented protective posture 4 (MOPP 4).

STANDARDS: Casualties are treated according to FM 4-25.11 and appropriate SOP(s). At MOPP 4 performance, degradation factors increase the time required to provide treatment and evacuation.

TASK STEPS AND PERFORMANCE MEASURES	GO	NO-GO
*1. The commander and leaders supervise first aid treatment of casualties (081-831-1055, 113-571-1022, 113-600-2001, 113-637-2001, and 805C-PAD-2060). They take the following actions: a. Implement treatment plan. b. Monitor treatment to ensure all casualties are treated. c. Direct employment of combat lifesavers to treat casualties. d. Monitor battlefield stress reduction and prevention procedures. e. Report casualties, as required. f. Coordinate with higher HQ for additional medical support. g. Coordinate replenishment of Class VIII supplies with supporting medical element according to SOPs. h. Direct distribution of Class VIII supplies according to SOPs. i. Enforce quality control procedures for Class VIII items issued to unit elements.		

TASK STEPS AND PERFORMANCE MEASURES	GO	NO-GO
2. Unit personnel perform first aid treatment (081-831-1003, 081-831-1005, 081-831-1007, 081-831-1008, 081-831-1025, 081-831-1026, 081-831-1032, 081-831-1033, 081-831-1034, 081-831-1044, 081-831-1045). They take the following actions:		
a. Evaluate casualties.		
b. Administer life-saving first aid treatment (cardiopulmonary resuscitation), if required.		
c. Control hemorrhage.		
d. Dress wounds.		
e. Splint suspected fractures.		
f. Provide first aid treatment to casualties with burns.		
g. Provide first aid treatment for environmental injuries.		
h. Provide first aid treatment for chemical casualties.		
i. Prevent shock.		
3. Unit medical personnel/combat lifesavers perform enhanced first aid treatment (081-831-0038, 081-831-0039, 081-831-1003, 081-831-1005, 081-831-1007, 081-831-1008, 081-831-1044, 081-831-1045, 081-833-0033, 081-833-0047, 081-833-0092). They take the following actions:		
a. Evaluate casualty for condition and type treatment needed.		
b. Measure casualty's vital signs.		
c. Initiate a field medical card.		
d. Insert oropharyngeal airway (J-Tube) in an unconscious casualty.		
d. Apply a splint to a fractured limb.		
e. Administer first aid to chemical agent casualties.		
f. Initiate an intravenous infusion for hypovolemic shock.		
g. Identify environmental injuries.		
h. Treat environmental injuries.		
i. Manage casualties with combat operational stress reactions.		
4. Unit medical personnel/combat lifesavers evacuate casualties to supporting medical element (081-831-0101, 081-831-1046, 081-833-0092). They take the following actions:		
a. Prepare casualties for evacuation.		

TASK STEPS AND PERFORMANCE MEASURES	GO	NO-GO
b. Identify litter team(s). c. Construct improvised litter from available material, as required. d. Secure casualty on litter. e. Employ appropriate manual carry if litter is not available. f. Transport casualty without causing further injury according to SOPs. * indicates a leader task step.		

SUPPORTING INDIVIDUAL TASKS

Task Number	Task Title
081-831-0038	Treat a Casualty for a Heat Injury
081-831-0039	Treat a Casualty for a Cold Injury
081-831-0101	Request Medical Evacuation
081-831-1003	Perform First Aid to Clear an Object Stuck in the Throat of a Conscious Casualty
081-831-1005	Perform First Aid to Prevent or Control Shock
081-831-1007	Perform First Aid for Burns
081-831-1008	Perform First Aid for Heat Injuries
081-831-1025	Perform First Aid for an Open Abdominal Wound
081-831-1026	Perform First Aid for an Open Chest Wound
081-831-1032	Perform First Aid for Bleeding of an Extremity
081-831-1033	Perform First Aid for an Open Head Wound
081-831-1034	Perform First Aid for a Suspected Fracture
081-831-1044	Perform First Aid for Nerve Agent Injury
081-831-1045	Perform First Aid for Cold Injuries
081-831-1046	Transport a Casualty
081-831-1055	Ensure Unit Combat Lifesaver Requirements Are Met
081-833-0033	Initiate an Intravenous Infusion
081-833-0047	Initiate Treatment for Hypovolemic Shock
081-833-0092	Transport a Casualty With a Suspected Spinal Injury
113-571-1022	Perform Voice Communications
113-600-2001	Communicate Via a Tactical Telephone
113-637-2001	Communicate Via a Tactical Radio in a Secure Net
805C-PAD-2060	Report Casualties

SUPPORTING COLLECTIVE TASKS

Task Number	Task Title
08-2-0004	Evacuate Casualties

08-2-0001   Conduct Battlefield Stress Reduction and Prevention
                   Procedures

SUPPORTING BATTLE/CREW DRILLS

Drill Number   **Drill Title**
07-3-D9507   Evacuate a Casualty (Dismounted and Mounted)

TASK: Evacuate Casualties (08-2-0004)

(FM 4-25.11)   (AR 190-8)   (AR 385-10)   (AR 600-8-1)
(ATTP 4-02)   (FM 4-02.7)   (TC 3-34.489)

CONDITIONS: Unit personnel are wounded and some may be chemically contaminated. Threat force contact has been broken. Unit defenses are reorganized and established. Casualties are evacuated from defensive positions to designated casualty collection points. Wounded enemy prisoners of war (EPW) casualties are evacuated to designated casualty collection points (CCPs) with appropriate security. The unit has analog and/or digital communications. Higher headquarters (HQ) operation order (OPORD) is available. Unit and higher HQ standing operating procedures (SOPs) are available. This task is performed under all environmental conditions. The unit may be subject to attack by threat forces, to include air; ground; chemical, biological, radiological, and nuclear (CBRN); or directed energy (DE) attack. Simplified collective protective equipment (SCPE) is on hand and/or field-expedient and natural shelters are available. Some iterations of this task should be performed in mission-oriented protective posture 4 (MOPP 4).

STANDARDS: Casualties are evacuated as soon as tactical situation permitted in according to FM 4-25.11, OPORD, appropriate SOPs, and provisions of the Geneva Conventions. At MOPP 4, performance degradation factors increase the time required to evacuate casualties.

TASK STEPS AND PERFORMANCE MEASURES	GO	NO-GO

*1. The commander and leaders supervise evacuation of casualties (113-571-1022, 113-600-2001, 113-637-2001). They take the following actions:
    a.   Monitor casualty evacuation operations for compliance with SOPs.
    b.   Identify casualty collection points.
    c.   Identify evacuation requirements.
    d.   Supervise preparation of casualties for evacuation.
    e.   Coordinate evacuation of casualties from unit area with the area defense command post (CP) according to SOPs.
    f.   Coordinate security requirements for the pick-up site with subelements and area defense CP.
    g. Disseminate evacuation information to unit personnel.		

TASK STEPS AND PERFORMANCE MEASURES	GO	NO-GO
h. Forward casualty feeder report and witness statements to the area defense CP according to SOPs. 2. Unit personnel prepare casualties for evacuation (101-92Y-0005, 113-571-1022, 113-600-2001, 113-637-2001, 805C-PAD-2060). They take the following actions: a. Provide first aid treatment to casualties (08-2-0003). b. Report casualties, as required. c. Collect classified documents such as signal operation instructions/signal supplemental instructions (SOI/SSI), maps, overlays, and key lists. d. Secure custody of organizational equipment according to SOPs. e. Forward casualty feeder reports to unit HQ according to SOPs. 3. Unit personnel evacuate casualties to casualty collection points using manual carries (081-831-1046). They take the following actions: a. Select type of manual carry appropriate to situation and injury. b. Evacuate casualty without causing further injury. 4. Unit personnel evacuate casualties to casualty collection points using litter carries (081-831-1046). They take the following actions: a. Identify litter team(s). b. Construct improvised litter from available material, as required. c. Secure casualty on litter. d. Evacuate casualty without causing further injury. 5. Unit personnel evacuate casualties to a medical treatment facility (MTF) using available vehicles (081-831-1046). They take the following actions: a. Load maximum number of casualties. b. Secure casualties in vehicle. c. Evacuate casualties without causing further injury.		

TASK STEPS AND PERFORMANCE MEASURES	GO	NO-GO

*6. The commander and leaders request aeromedical evacuation (081-831-0101, 113-571-1022, 113-600-2001, 113-637-2001, 301-371-1050). They take the following actions:
    a.   Transmit request according to OPORD and SOPs.
    b.   Select landing site, which provides sufficient space for helicopter hover, landing, and take-off.
    c. Supervise removal of all dangerous objects likely to be blown about before aircraft arrival.
    d.   Supervise security of landing site according to the SOPs.
    e.   Ensure landing zone (LZ) is appropriately marked (light sets, smoke, and so forth) according to SOPs, if required.
7. Unit personnel assist in loading ambulance (081-831-1046). They take the following actions:
    a.   Employ proper carrying and loading techniques.
    b.   Load casualties in the sequence directed by crew.
    c.   Load casualties without causing unnecessary discomfort.
    d.   Employ safety procedures according to SOPs.
    e. Employ environmental protection procedures according to SOPs.
8. Unit personnel evacuate chemically contaminated casualties (031-503-1035, 081-831-1046). They take the following actions:
    a.   Assume MOPP 4.
    b.   Mark contaminated casualties according to SOPs.
    c.   Notify supporting MTF that contaminated casualties are en route to their location.
    d.   Evacuate casualties directly to a designated decontamination and treatment station.
    e.   Protect casualties from further contamination during evacuation.
9. Unit personnel evacuate EPW casualties (081-831-1046, 181-105-1001). They take the following actions:		

TASK STEPS AND PERFORMANCE MEASURES	GO	NO-GO

a.   Maintain security of EPW casualties according to SOPs.
    b.   Search EPW casualties for weapons and ordnance before evacuation.
    c.   Evacuate EPW casualties according to the provisions of the Geneva Conventions and SOPs.
* indicates a leader task step		

SUPPORTING INDIVIDUAL TASKS

Task Number	Task Title
031-503-1035	Protect Yourself From Chemical and Biological (CB) Contamination Using Your Assigned Protective Mask
081-831-0101	Request Medical Evacuation
081-831-1046	Transport a Casualty
101-92Y-0005	Enforce Compliance With Property Accountability Policies
113-571-1022	Perform Voice Communications
113-600-2001	Communicate Via a Tactical Telephone
113-637-2001	Communicate Via a Tactical Radio in a Secure Net
181-105-1001	Comply With the Law of War and the Geneva and Hague Conventions
301-371-1050	Implement Operations Security (OPSEC) Measures
805C-PAD-2060	Report Casualties

SUPPORTING COLLECTIVE TASKS

Task Number	Task Title
08-2-0003	Treat Casualties
08-2-0001	Conduct Battlefield Stress Reduction and Prevention Procedures

SUPPORTING BATTLE/CREW DRILLS

Drill Number	Drill Title
07-3-D9507	Evacuate a Casualty (Dismounted and Mounted)

This page intentionally left blank.

Chapter 3

Supporting Battle/Crew Drills

This chapter provides the platoon leader an example of the platoon collective task with supporting battle and or crew drills. Also provided are the drill T&EOs that can be used to train or evaluate a single drill. Several drill T&EOs may be used by an observer controller as an evaluation outline or by a platoon leader as a training outline.

BATTLE/CREW DRILLS

3-1.   The collective task to drill table is an example developed by the DOTD, MCoE. (See Table 3-1.) This table can be used by platoon leaders and unit leaders as an example to create their own unique collective task to drill crosswalk.

3-2.   The T&EOs for the drills shown in the example matrix are displayed using the T&EO outline format. (See Table 3-1.) For more information about battle and or crew drills the platoon may be expected to perform, see DTMS.

Table 3-1. Collective task to drill table

Collective Task Number and Title	
Supporting Battle Drill Number and Title	
07-2-9014	*Occupy an Assembly Area (Platoon-Company)*
	07-3-D9501, React to Contact (Visual, IED, Direct Fire [includes RPG])
	17-3-D8008, React to an Obstacle
07-2-6063	*Maintain Operations Security (Platoon-Company)*
	07-3-D9501, React to Contact (Visual, IED, Direct Fire [includes RPG])
	07-4-D9272, Perform Direct Lay of a Mortar
	17-3-D8008, React to an Obstacle

Table 3-1. Collective task to drill table (continued)

Collective Task Number and Title	
Supporting Battle Drill Number and Title	
07-2-1234	**Conduct an Airborne Assault (Platoon-Company)**
	07-4-D9272, Perform Direct Lay of a Mortar
	07-3-D9501, React to Contact (Visual, IED, Direct Fire [includes RPG])
07-2-1495	**Conduct an Air Assault (Platoon-Company)**
	17-3-D8008, React to an Obstacle
	07-3-D9505, Break Contact
	05-3-D0016, Conduct the 5Cs
07-2-5081	**Conduct Troop-Leading Procedures (Platoon-Company)**
	07-3-D9501, React to Contact (Visual, IED, Direct Fire [includes RPG])
	07-3-D9504, React to Indirect Fire
07-2-5027	**Conduct Consolidation and Reorganization (Platoon-Company)**
	05-3-D0016, Conduct the 5Cs
	07-3-D9507, Evacuate a Casualty (Dismounted and Mounted)
07-2-1342	**Conduct Tactical Movement (Platoon-Company)**
	07-3-D9501, React to Contact (Visual, IED, Direct Fire [includes RPG])
	07-4-D9316, Perform Direct Alignment of a Mortar
07-3-1351	**Occupy a Mortar Firing Position (Section-Platoon)**
	07-4-D9267, Place 81-mm Mortar into Action
	07-4-D9280, Mount the Mortar Carrier

Table 3-1. Collective task to drill table (continued)

Collective Task Number and Title	
Supporting Battle Drill Number and Title	
07-3-2045	**Reconnoiter a Mortar Firing Position (Section-Platoon)**
	07-3-D9501, React to Contact (Visual, IED, Direct Fire [includes RPG])
	07-4-D9314, Perform Hasty Lay of a 120-mm Mortar for Hipshoot
	07-3-D9504, React to Indirect Fire
07-3-3099	**Fire a Mortar Priority Target Mission (Section-Platoon)**
	07-4-D9339, Fire the Mortar
	07-4-D9277, Remove a Misfire from 81-mm Mortar
07-3-9013	**Conduct Action on Contact**
	07-3-D9501, React to Contact (Visual, IED, Direct Fire [includes RPG])
	17-3-D8004, React to Air Attack Drill
07-3-3054	**Fire a Mortar Adjust Fire Mission (Section-Platoon)**
	07-4-D9339, Fire the Mortar
	07-4-D9275, Lay 81mm Mortar for Large Deflection and Elevation Change
07-3-3072	**Fire a Mortar Fire for Effect Mission (Section-Platoon)**
	07-4-D9339, Fire the Mortar
63-2-4546	**Conduct Logistics Package (LOGPAC) Support**
	07-3-D9501, React to Contact (Visual, IED, Direct Fire [includes RPG])
03-2-9224	**Conduct Operational Decontamination**
	03-3-D0035, React to a Chemical Attack

Table 3-1. Collective task to drill table (continued)

Collective Task Number and Title	
Supporting Battle Drill Number and Title	
08-2-0003	*Treat Casualties*
	07-3-D9507, Evacuate 07-4-D9339 Fire the Mortar
	07-4-D9275, Lay 81mm Mortar for Large Deflection and Elevation Change
	03-3-D0035 React to a Chemical Attack
	07-3-D9507 a Casualty (Dismounted and Mounted)
	07-3-D9501 React to Contact (Visual, IED, Direct Fire [includes RPG])
	07-4-D9339 Fire the Mortar
08-2-0004	*Evacuate Casualties*
	07-3-D9507, Evacuate a Casualty (Dismounted and Mounted)

TASK: React to Contact (Visual, IED, Direct Fire [includes RPG]) (07-3-D9501)

CONDITIONS: Visual (dismounted/mounted). The unit is stationary or moves, conducting operations. Visual contact is made with the enemy. Mounted. The unit is stationary or moves, conducting operations. Visual contact is made with the enemy. Improvised explosive device (IED) (dismounted/mounted). The unit is stationary or moves, conducting operations. The unit identifies and confirms an IED or one is detonated. Direct fire dismounted/mounted. The unit is stationary or moves, conducting operations. The enemy initiates contact with a direct fire weapon.

CUE: This drill begins when visual contact, direct fire or an IED is identified or detonated.

STANDARDS: Visual (dismounted). The unit destroys the enemy with a hasty ambush or an immediate assault through the enemy position. Visual (mounted). Based on the composition of the mounted unit, the unit either suppresses and reports the enemy position and continues its mission, or suppresses the enemy position for a follow-on assault to destroy them. IED (dismounted/mounted). The unit takes immediate action by using the 5Cs procedure (confirm, clear, call, cordon, check, and control). Direct fire (dismounted/mounted). The unit immediately returns well-aimed fire and seeks cover. The unit leader reports the contact to higher headquarters (HQ).

TASK STEPS AND PERFORMANCE MEASURES
1. Visual dismounted.
    a. Hasty ambush. Unit leaders take the following actions:
     (1) Determine that the unit has not been seen by the enemy.
     (2) Signal Soldiers to occupy best available firing positions.
     (3) Initiate the ambush with the most casualty-producing weapon available, immediately followed by a sustained well-aimed volume of effective fire.
     (4) If the unit is prematurely detected, the Soldier(s) aware of the detection initiates the ambush.
     (5) Ensure the unit destroys the enemy or forces them to withdraw.
     (6) Report the contact to higher HQ.
    b. Immediate assault.
     (1) The unit and the enemy simultaneously detect each other at close range.

TASK STEPS AND PERFORMANCE MEASURES

(2) All soldiers who see the enemy engage and announce "contact" with a clock direction and distance to enemy, (example, "contact three o'clock, 100 meters"). Unit personnel take the following actions:

(3) Elements in contact immediately assault the enemy using fire and movement.

(4) The unit destroys the enemy or forces them to withdraw

(5) The unit leader reports the contact to higher headquarters.

2. Visual mounted. Unit personnel take the following actions:

a. The Soldier who spots the enemy announces the contact.

b. The element in contact immediately suppresses the enemy.

c. The vehicle commander of the vehicle in contact sends contact report over the radio.

d. The unit maneuvers on the enemy or continues to move.

e. Vehicle gunners fix and suppress the enemy positions.

f. The unit leader reports the contact to higher HQ.

3. IED dismounted/mounted. Unit personnel take the following actions:

a. React to a suspected or known IED prior to detonation by using the 5Cs.

b. Unit determines if there is a requirement for explosive ordnance disposal (EOD), while maintaining as safe a distance as possible and 360 security, Unit "confirms" the presence of an IED by using all available optics to identify any wires, antennas, detcord, or parts of exposed ordinance. Take the following actions:

(1) Conduct surveillance from a safe distance.

(2) Observe the immediate surroundings for suspicious activities.

(3) Requests EOD if the need is determined.

c. Unit "clears" all personnel from the area a safe distance to protect them from a potential second IED.

d. Unit "cordons" off the area, directs personnel out of the danger area, prevents all military or civilian traffic from passing and allows entry only to authorized personnel. They take the following actions:

(1) Direct people out of the 300-meter minimum danger area.

(2) Identify and clears an area for an incident control point (ICP).

(3) Occupy positions and continuously secure the area.

e. Unit "checks" the immediate area for secondary/tertiary devices around the incident control point (ICP) and cordon using the 5/25 meter checks.

f. Unit "controls" the area inside the cordon to ensure only authorized access

TASK STEPS AND PERFORMANCE MEASURES

g. Unit continuously scans the area for suspicious activity. They take the following actions:

(1)   Identify potential enemy observation, vantage, or ambush points.

(2)   Maintain visual observation on the IED to ensure the device is not tampered with.

4.   Direct fire dismounted. (See Figure 1.) Unit personnel take the following actions:

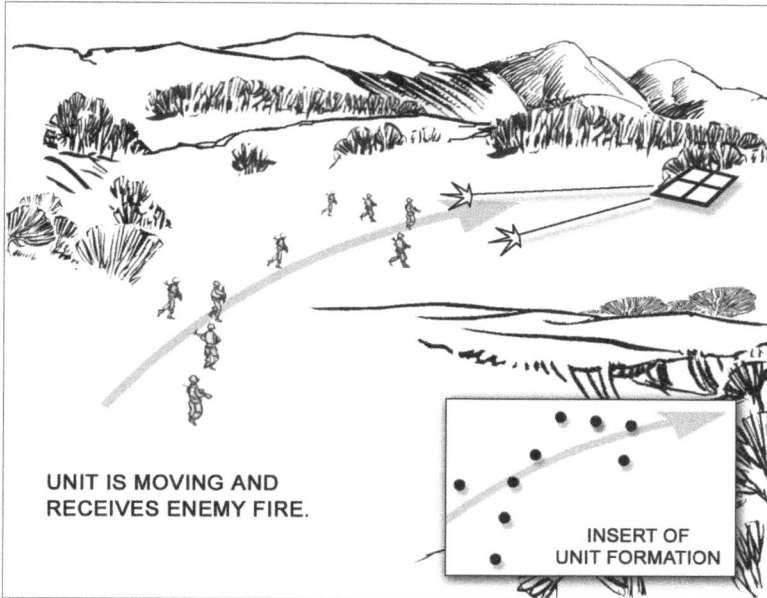

UNIT IS MOVING AND
RECEIVES ENEMY FIRE.

INSERT OF
UNIT FORMATION

Figure 1. React to contact, direct fire (dismounted)

a. Soldiers under direct fire immediately return fire and seek the nearest covered positions. They call out distance and direction of direct fire. (See Figure 2.)

TASK STEPS AND PERFORMANCE MEASURES

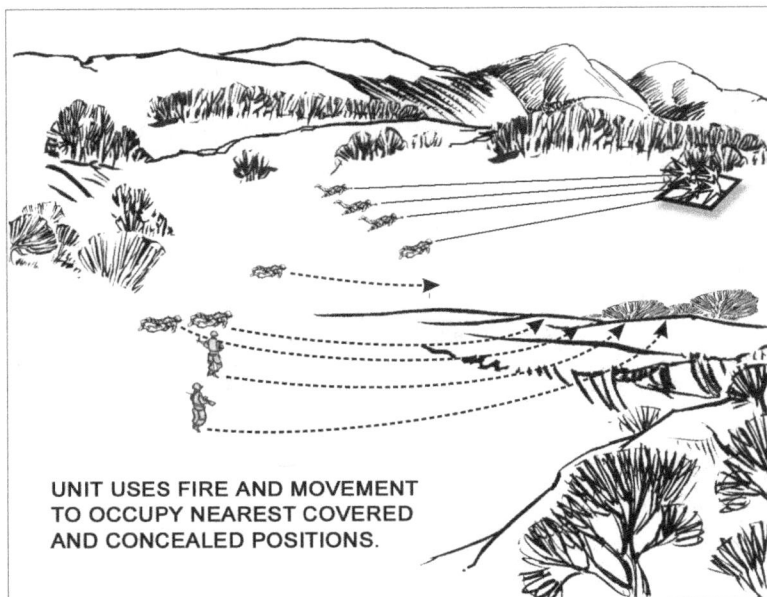

UNIT USES FIRE AND MOVEMENT
TO OCCUPY NEAREST COVERED
AND CONCEALED POSITIONS.

Figure 2. React to contact, direct fire (dismounted)

b. Element leaders locate and engage known or suspected enemy positions with well-aimed fire and pass information to the unit leader.

c. Element leaders control their Soldier's fire by:

  (1) Marking targets with lasers.

  (2) Marking the intended target with tracers or M203 rounds.

d. Soldiers maintain contact (visually or orally) with the Soldiers on their left or right.

e. Soldiers maintain contact with their team leader and relay the location of enemy positions. (See Figure 3.)

TASK STEPS AND PERFORMANCE MEASURES

UNIT IS IN POSITION ENGAGING
ENEMY WITH WELL-AIMED FIRES.

Figure 3. React to contact, direct fire (dismounted)

f. Element leaders (visually or orally) check the status of their Soldiers.

g. Element leaders maintain contact with the unit leader.

h. Unit leader reports the contact to higher headquarters.

5.　　Direct fire mounted. Unit personnel take the following actions:

a. If moving as part of a logistics patrol, vehicle gunners immediately suppress enemy positions and continue to move.

b. Vehicle commanders direct their drivers to accelerate safely through the engagement area.

c. If moving as part of a combat patrol, vehicle gunners suppress and fix the enemy allowing others to maneuver against and destroy the enemy.

d. Leaders (visually or orally) check the status of their Soldiers and vehicles.

e. Unit leader reports the contact to higher HQ.

SUPPORTING PRODUCTS

Product ID	Product Name
FM 3-21.8	The Infantry Rifle Platoon and Squad
FM 3-21.75	Warrior Ethos and Soldier Combat Skills
ATTP 3-21.9	SBCT Infantry Rifle Platoon and Squad

SUPPORTING INDIVIDUAL TASKS

Task Number	Task Title
071-030-0004	Engage Targets with an MK 19 Grenade Machine Gun
071-054-0004	Engage Targets with an M136 Launcher
071-325-4407	Employ Hand Grenades
071-311-2130	Engage Targets with an M203 Grenade Launcher
071-010-0006	Engage Targets with an M249 Machine Gun
071-025-0007	Engage Targets with an M240B Machine Gun
071-100-0030	Engage Targets with an M16-Series Rifle/M4-Series Carbine
071-326-0502	Move Under Direct Fire
071-326-5611	Conduct the Maneuver of a Squad
071-326-5630	Conduct Movement Techniques by a Platoon
071-121-4080	Send a Spot Report (SPOTREP)
061-283-1011	Engage Targets with Indirect Fires
113-571-1022	Perform Voice Communications

SUPPORTED COLLECTIVE TASKS

Task Number	Task Title
07-2-1090	Conduct a Movement to Contact (Platoon-Company)
07-2-1450	Secure Routes (Platoon-Company)
07-2-9002	Conduct a Bypass (Platoon-Company)
07-2-9009	Conduct a Withdrawal (Platoon-Company)

TASK: Place 81-mm Mortar into Action (07-4-D9267)
    ATTP 3-21.90   FM 3-22.90

CONDITIONS: The mortar unit has displaced to a new firing position and receives a call for fire. The unit leader commands, "ACTION."

CUE: The unit leader initiates drill by giving the order: "ACTION."

STANDARDS: The unit is dismounted in a firing position. The unit leader places an aiming post at location mortar to be laid and commands "ACTION". The unit places the mortar into action and is ready to engage the enemy within 90 seconds of the command ACTION (day or night).

1.   The unit ensures the following mortar mounting and safety requirements are met:
    a.   Check mask and overhead clearance.
    b.   Open end of the socket cap is pointing in the direction of fire with firing pin recess facing upwards.
    c.   The leg-locking hand-wheel is tightened.
    d.   Bubbles on sight are level.
    e.   Sights have correct setting for deflection and elevation into dove tale slot.
    f.   The baseplate is positioned correctly in relation to the baseplate stake.
    g.   Locking collar lock and adjusting the lower stop end.
    h.   Buffer carrier is no more than two turns to the left or right of the center turret.
2.   Mortar is mounted at the designated location.
3.   The gunner announces: "UP. "

TASK STEPS AND PERFORMANCE MEASURES

1.   Squad leader picks up and places the sight case and two aiming posts at the exact position where the mortar is to be mounted. (071-074-0008)
2.   Squad leader points in the direction of fire and commands, "ACTION." At night, the squad leader may use a flashlight and a compass to direct the guns in the proper direction of fire.
3.   Gunner places the outer edge of the baseplate against the baseplate stake. He aligns the left edge of the cutout portion of the baseplate with the left edge of the baseplate stake.
4.   Gunner rotates the socket so that the open end points in the direction of fire.
5.   Assistant gunner places his left hand on the traversing handwheel and his right hand on the sight slot, and lifts the bipod.
6.   Assistant gunner moves around to the front and faces the baseplate.

TASK STEPS AND PERFORMANCE MEASURES

7. Ammunition bearer lowers the breech plug into the rotating socket and rotates the barrel a one-quarter turn to lock it to the baseplate. He ensures that the firing pin recess is facing upward.

8. Ammunition bearer stands to the right rear of the baseplate and supports the barrel until the mount is fitted.

9. Assistant gunner lifts the mount and stands it on its elevating leg assembly so that the elevation handwheel is to the right and the fixed leg assembly is to the left.

NOTE: The assistant gunner prepares the mount for operation at the rear of the mortar before moving it forward.

10. Assistant gunner releases the securing strap, loosens the leg-locking handwheel, and lowers the fixed leg assembly until the locating catch engages in the recess.

11. Assistant gunner tightens the leg-locking handwheel by hand, ensuring the teeth on either side are correctly meshed.

12. Assistant gunner exposes eight inches (200-mm) of elevation shaft, leaving the elevation handwheel unfolded.

13. Assistant gunner opens the cross-leveling mechanism and traversing handwheel and centers the buffer carrier.

14. Assistant gunner places the mount feet two to three feet in front of the baseplate on the ground.

15. Assistant gunner positions the lower barrel clamp against the lower stop band on the barrel and secures the upper barrel clamp.

16. Assistant gunner ensures that the ball-shaped end of the locking rod is secured in its recess by the locking latch and that lock barrel assembly is secure.

17. Gunner takes the sight out of the sight case, inserts it into the sight bracket on the yoke assembly, and sets the mounting data on the sight: M64A1, deflection 3,200 mils, elevation 0800 mils.

NOTE: Sight should be stored at an elevation of 1100, but for sight magnification during this drill set sight at an elevation of 0800.

18. Gunner centers the elevation level bubble, centers the cross-level bubble, and rechecks the elevation bubble.

19. Gunner announces, UP.

SUPPORTING INDIVIDUAL TASKS

Task Number	Task Title
071-074-0006	Manipulate a 60-mm, 81-mm, or 120-mm Mortar for Traversing and/or Searching Fire
091-257-0002	Conduct Preventive Maintenance Checks and Services (PMCS)
071-074-0004	Engage Targets with a 60-mm, 81-mm, or 120-mm Mortar Using Direct Lay
071-074-0042	Lay a Mortar Using Direct Alignment
071-321-3905	Prepare 81-mm Mortar Ammunition for Firing
091-91F-2015	Correct Malfunctions of 81-mm M252 Mortar

SUPPORTING COLLECTIVE TASKS

Task Number	Task Title
07-3-1351	Occupy a Mortar Firing Position (Section-Platoon)

TASK: Remove a Misfire from 81-mm Mortar (07-4-D9277)

ATTP 3-21.90   FM 3-22.90

CONDITIONS: The unit is in a firing position. A loaded mortar fails to fire. The first crewmember that notices the failure of the mortar round to fire announces, "MISFIRE."

CUE: "This drill begins when a crew member notices the failure and announces, "MISFIRE."

STANDARDS: All required actions must be performed in sequence without error. The unit leader supervises/confirms actions of the crew using TM 9-1015-249-10 with appropriate changes during the entire misfire procedure. All live fire precautions must be adhered to.

TASK STEPS AND PERFORMANCE MEASURES

1.   First crewmember that notices the failure to fire announces, "MISFIRE."

2.   Entire crew stays with the mortar.

NOTE: During peacetime live-fire training, all squad members, except the gunner, move at least 50 meters to the rear of the mortar.

3.   Gunner moves to the rear of the mortar and kicks the base of the barrel several times.

NOTE: During peacetime live-fire training, if the round does not fire, the gunner joins the unit and waits one minute before returning to the mortar position.

4.   Gunner checks the barrel for heat, bare-handed, with his fingertips to determine if it is cool enough to handle.

5.   If the barrel is too hot to handle, the gunner employs field-expedient measures to cool the barrel and then rechecks for heat.

6.   As soon as the barrel is cool enough to handle, gunner informs the unit leader.

NOTE: During peacetime live-fire training, the crew returns to the mortar position.

7.   Gunner removes the sight-unit and places it in the sight case.

8.   Gunner removes the firing pin.

NOTE: Gunner depresses elevation if necessary, then handcrank one turn at a time until the firing pin can be removed.

9.   Gunner unlocks the barrel clamp and rotates the barrel, unlocking the breech plug from the socket of the baseplate. He then relocks the barrel clamp.

10. Gunner grasps both ends of the traversing gear assembly and supports the mortar.

TASK STEPS AND PERFORMANCE MEASURES

11. Assistant gunner places his right hand (palm up) under the blast attenuator device and his left hand (palm down) on top of the blast attenuator device.

12. Ammunition bearer places both hands on the cooling fins under the barrel and raises the barrel to the horizontal position. Gunner is supporting the mortar.

NOTE: Once barrel is horizontal it should never be lowered until the round is removed.

13. Assistant gunner slips his thumbs over the edge of the muzzle to stop the round as it slides out.

14. Ammunition bearer raises the base of the barrel until the round slides out.

15. Assistant gunner catches the round.

16. Ammunition bearer shakes the barrel to dislodge any remnants from the last round fired. He lowers the barrel into the rotating socket of the baseplate.

17. Assistant gunner removes the round and passes it to the ammunition bearer.

18. Ammunition bearer inspects the round for the cause of the misfire. If the primer of the ignition cartridge is dented, the ammunition bearer tries to replace the safety wire and puts the round in a marked safe location.

19. Gunner replaces the firing pin.

20. Assistant gunner swabs the bore.

21. Gunner replaces the sight and re-lays on the last data given by the squad leader. (071-086-0003)

22. Ammunition bearer examines the round to determine the cause of the firing malfunction.

NOTE: Squad leader takes corrective action based on identified cause of misfire.

NOTE: Squad leader identifies the appropriate Dud pit emplacement.

SUPPORTING INDIVIDUAL TASKS

Task Number	Task Title
061-283-6003	Adjust Indirect Fire
071-074-0006	Manipulate a 60-mm, 81-mm, or 120-mm Mortar for Traversing and/or Searching Fire
091-257-0002	Conduct Preventive Maintenance Checks and Services (PMCS)
071-074-0004	Engage Targets with a 60-mm, 81-mm, or 120-mm Mortar Using Direct Lay

SUPPORTING INDIVIDUAL TASKS

Task Number	Task Title
071-074-0036	Adjust Mortar Fire Using Direct Alignment
071-074-0037	Fire a Mortar
071-074-0042	Lay a Mortar Using Direct Alignment
071-321-3905	Prepare 81-mm Mortar Ammunition for Firing
091-91F-2015	Correct Malfunctions of 81-mm M252 Mortar

SUPPORTING COLLECTIVE TASKS

Task Number	Task Title
07-3-3072	Fire a Mortar Fire for Effect Mission (Section-Platoon)
07-3-3054	Fire a Mortar Adjust Fire Mission (Section-Platoon)
07-3-3099	Fire a Mortar Priority Target Mission (Section-Platoon)

TASK: Break Contact (07-3-D9505)

CONDITIONS: (Dismounted/Mounted) - The unit is stationary or moving, conducting operations. All or part of the unit is receiving enemy direct fire.

CUE: The unit leader initiates drill by giving the order to BREAK CONTACT.

STANDARDS: (Dismounted/Mounted) - The unit returns fire. A leader identifies the enemy as a superior force, and makes the decision to break contact. The unit breaks contact using fire and movement. The unit continues to move until the enemy cannot observe or place effective fire on them. The unit leader reports the contact to higher headquarters.

TASK STEPS AND PERFORMANCE MEASURES

1.   Dismounted

a. The unit leader designates an element to suppress the enemy with direct fire as the base-of-fire element.

b. The unit leader orders distance, direction, a terrain feature, or last rally point for the movement of the first element.

c. The unit leader calls for and adjusts indirect fire to suppress the enemy positions.

d. The base-of-fire element continues to suppress the enemy. (See Figure 1.)

TASK STEPS AND PERFORMANCE MEASURES

Figure 1. Break contact (dismounted)

e. The bounding element uses the terrain and/or smoke to conceal its movement and bounds to an overwatch position.

f. The bounding element occupies their overwatch position and suppresses the enemy with "well-aimed fire." (See Figure 2.)

TASK STEPS AND PERFORMANCE MEASURES

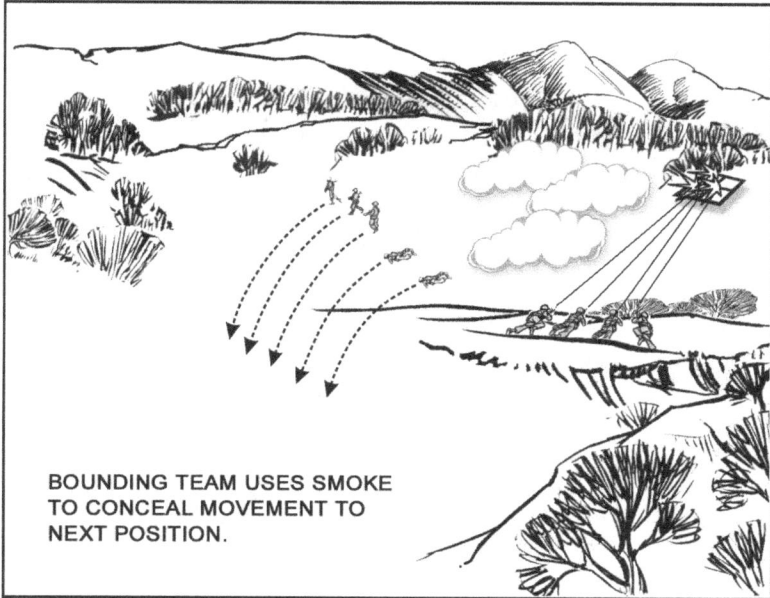

BOUNDING TEAM USES SMOKE
TO CONCEAL MOVEMENT TO
NEXT POSITION.

Figure 2. Break contact (dismounted) (continued)

g. The base-of-fire element moves to its next covered and concealed position. (Based on the terrain and volume and accuracy of the enemy's fire, the moving element may need to use fire and movement techniques). (See Figure 3.)

TASK STEPS AND PERFORMANCE MEASURES

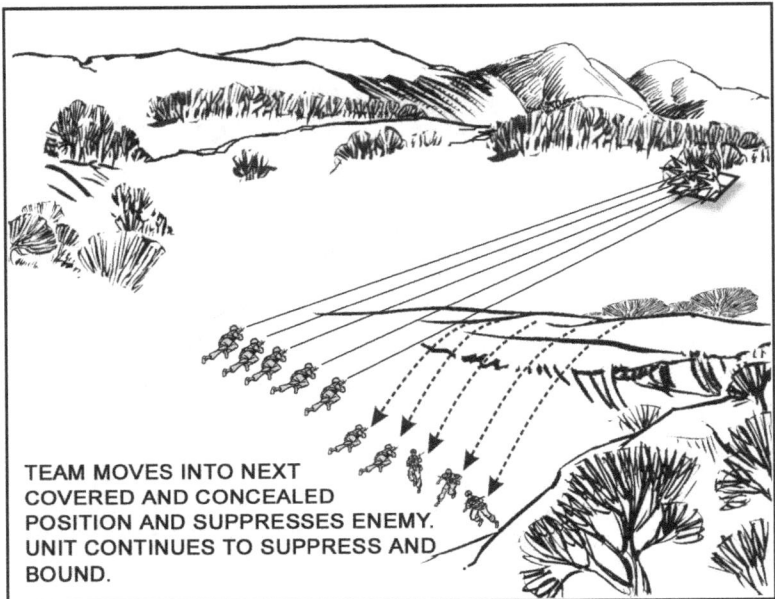

TEAM MOVES INTO NEXT
COVERED AND CONCEALED
POSITION AND SUPPRESSES ENEMY.
UNIT CONTINUES TO SUPPRESS AND
BOUND.

Figure 3. Break contact (dismounted) (continued)

h. The unit continues to suppress the enemy and bound until it is no longer in contact with enemy.

i. The unit leader reports the contact to higher headquarters.

2. Mounted -

a. The unit leader directs the vehicles in contact to place "well-aimed" suppressive fire on the enemy positions.

b. The unit leader orders distance, direction, a terrain feature, or last objective rally point over the radio for the movement of the first section.

c. The unit leader calls for and adjusts indirect fire to suppress the enemy positions.

d. Gunners in the base-of-fire vehicles continue to engage the enemy. They attempt to gain fire superiority to support the bound of the moving section.

e. The bounding section moves to assume the overwatch position.

(1) The section uses the terrain and/or smoke to mask movement.

(2) Vehicle gunners and mounted Soldiers continue to suppress the enemy.

f. The unit continues to suppress the enemy and bounds until it is no longer receiving enemy fire.

g.   The unit leader reports the contact to higher HQ.

SUPPORTING PRODUCTS

Product ID	Product Name
FM 3-21.8	The Infantry Rifle Platoon and Squad
FM 3-21.75	Warrior Ethos and Soldier Combat Skills
ATTP 3-21.9	SBCT Infantry Rifle Platoon and Squad

SUPPORTING INDIVIDUAL TASKS

Task Number	Task Title
071-030-0004	Engage Targets with an MK 19 Grenade Machine Gun
071-054-0004	Engage Targets with an M136 Launcher
071-325-4407	Employ Hand Grenades
071-311-2130	Engage Targets with an M203 Grenade Launcher
071-010-0006	Engage Targets with an M249 Machine Gun
071-025-0007	Engage Targets with an M240B Machine Gun
071-100-0030	Engage Targets with an M16-Series Rifle/M4-Series Carbine
071-326-0502	Move Under Direct Fire
071-326-5611	Conduct the Maneuver of a Squad
071-326-5630	Conduct Movement Techniques by a Platoon
071-121-4080	Send a Spot Report (SPOTREP)
061-283-1011	Engage Targets with Indirect Fires
113-571-1022	Perform Voice Communications

SUPPORTED COLLECTIVE TASKS

Task Number	Task Title
07-2-1090	Conduct a Movement to Contact (Platoon-Company)
07-2-1450	Secure Routes (Platoon-Company)
07-2-9002	Conduct a Bypass (Platoon-Company)
07-2-9009	Conduct a Withdrawal (Platoon-Company)

TASK: React to an Obstacle (17-3-D8008)

CONDITIONS: The platoon is conducting tactical operations as part of a higher unit and has communication with the commander. The platoon or a section/squad makes contact with an obstacle. The platoon may or may not have countermine equipment. Enemy contact is possible. Some iterations of this task should be conducted in mission-oriented protective posture 4 (MOPP 4) and under conditions of limited visibility.

CUE: Any Soldier gives an oral or visual signal they are in contact with an obstacle.

STANDARDS: The platoon identifies the obstacle, deploys as applicable to avoid decisive engagement of the entire platoon, and alerts the higher unit of obstacle contact and location. Once the obstacle is breached or bypassed, the platoon remains prepared to continue the unit mission. No friendly unit suffers casualties or equipment damage as a result of fratricide.

TASK STEPS AND PERFORMANCE MEASURES

1.   If applicable, section identifying the obstacle alerts the platoon with a contact report.

2.   In close direct fire contact situations, platoon takes immediate protective actions.

   a.   The platoon leader (PL) directs the platoon to deploy to a covered and concealed location.

   b.   As applicable, element in contact employs onboard smoke grenades and direct fire to obscure and suppress the enemy forces overwatching the obstacle.

3.   In out of contact situations (platoon identifies obstacle from a position of advantage), platoon takes immediate protective actions.

   a.   PL directs the platoon to deploy to a covered and concealed location.

   b.   Element in visual contact with obstacle establishes an overwatch position.

   c.   As applicable, employs direct fire and/or indirect fire to obscure and suppress the enemy forces overwatching the obstacle.

4.   PL/platoon sergeant takes actions to develop the situation and report to the commander.

   a.   Sends contact report to higher commander.

   b.   Develops the situation by section/squad (maneuver) to determine location, composition, and disposition of enemy forces overwatching the obstacle.

TASK STEPS AND PERFORMANCE MEASURES
(1)   Directs one section/squad to establish a suitable overwatch position to allow platoon to continue to develop the situation.

(2)   Directs the other section/squad to perform reconnaissance of the obstacle to determine composition of the obstacle and to locate a bypass.

NOTE: Reconnaissance may be performed mounted or dismounted.

c.   Sends obstacle report to higher commander describing type, width, length, effect, and location of the obstacle.

d.   Sends updated situation reports to the commander as necessary.

5.   If a bypass is possible, PL reports the location of the bypass to the higher commander and recommends bypassing the obstacle.

NOTE: Once ordered to bypass, the platoon executes steps to bypass the obstacle. (Refer to task 07-2-9002, Conduct a Bypass) (Platoon-Company)

6.   If a bypass is not possible, PL reports to higher commander and recommends, based on obstacle composition, a point of breach and either platoon-level reduction or a higher-level breach.

NOTE: If ordered to reduce the obstacle, the platoon executes steps of breach force operations. (Refer to task 17-2-3070, Breach an Obstacle [Platoon-Company])

SUPPORTING PRODUCTS

Product ID	Product Name
FM 3-20.15	Tank Platoon
FM 3-90.1	Tank and Mechanized Infantry Company Team

SUPPORTING INDIVIDUAL TASKS

Task Number	Task Title
171-121-3009	Control Techniques of Movement
171-121-4009	Conduct Scout Platoon Actions on Contact
171-121-4010	Conduct Tank Platoon Actions on Contact
171-121-4017	Supervise Tank Platoon Formations and Drills
171-121-4038	Supervise Local Security
171-121-4045	Conduct Troop Leading Procedures
171-121-4059	Conduct an Armor in-Stride Breach of a Minefield
171-121-4068	Conduct a Reconnaissance by Fire
071-100-0030	Engage Targets with an M16-Series Rifle/M4-Series Carbine
071-325-4407	Employ Hand Grenades
071-326-0501	Move as a Member of a Fire Team

SUPPORTING INDIVIDUAL TASKS

Task Number	Task Title
071-326-0608	Use Visual Signaling Techniques
071-326-0503	Move Over, Through, or Around Obstacles (Except Minefields)

SUPPORTED COLLECTIVE TASK

Task Number	Task Title
07-2-1090	Conduct a Movement to Contact (Platoon-Company)
17-2-3070	Breach an Obstacle (Platoon-Company)
17-2-4000	Conduct Route Reconnaissance (Platoon-Company)

TASK: Perform Direct Alignment of a Mortar (07-4-D9316)

ATTP 3-21.90   FM 3-22.90

CONDITIONS: The unit leader, acting as fire direction control (FDC) and forward observer (FO), commands the unit to engage a target visible only to him. He commands, "UNIT, DIRECT ALIGNMENT."

CUE: The unit leader initiates drill by giving the order, "UNIT, DIRECT ALIGNMENT."

STANDARDS: The unit leader takes up a vantage point visible to the gunner. The mortar is laid. The mortar is within four turns of center-of-traverse (120-mm) or two turns of center-of-traverse (81-mm/60-mm). The mortar is cross-leveled and all bubbles are centered. The vertical crossline on the sight unit is laid on the aiming device such as the "T" bar or aiming post.

Once the gunner has mortar laid, he announces "UP" and signals the unit leader I AM READY, if necessary.

TASK STEPS AND PERFORMANCE MEASURES

1.   Unit leader commands, "UNIT, DIRECT ALIGNMENT."
2.   Unit leader moves to a vantage point (OP) within 100 meters of the mortar firing position and on the gun target line.
3.   Unit leader places an aiming device at his vantage point visible to the gunner.
NOTE: The unit leader's position can be to the front of or to the rear of the mortar, as long as he is within 100 meters of the gun and on the gun target line. Once the unit leader has placed the aiming device, he can move to the left, right, or rear of the gun position, remaining within 100 meter of the mortar.
4.   Unit leader positions himself so that he can sight along the horizontal portion of the aiming device to the target.
5.   Unit leader reverses his position and sights back along the horizontal portion of the aiming device toward the mortar position without disturbing the aiming device.
NOTE: If the unit leader selects a position to the rear of the mortar, he does not reverse his position.
6.   Unit leader signals to the gunner by voice command or arm-and-hand signals to mount the mortar so that it is aligned with the aiming device.
7.   Unit places the mortar into action. The gunner aligns the vertical crossline of the sight on the unit leader's aiming device instead of a direction stake.
8.   Gunner ensures the traversing mechanism is centered. Ensures the sight has a deflection of 3,200 mils and an elevation of 1,100 mils.

TASK STEPS AND PERFORMANCE MEASURES

NOTE: If the selected unit leader's position was to the rear of the mortar, then the deflection indexed will be 0000 miles (6,400).

9. Unit leader prepares and issues the initial fire command to include shell and fuze, method of fire, charge, and elevation--for example, "HE QUICK, ONE ROUND, CHARGE THREE, ELEVATION ONE ONE NINE NINE."

10. Gunner places the announced elevation on the sight.

11. Ammunition bearer prepares ammunition according to the announced shell and fuze, method of fire, and charge.

12. Gunner levels the mortar so that all bubbles are centered, and the vertical crossline of the sight is laid on the device.

13. Gunner announces, "UP," and signals to the unit leader I AM READY, if necessary.

14. Unit loads and fire. When the round is fired, the unit leader will observe the impact of the round.

15. Once the round impacts, if any deviation corrections are needed, the unit leader will estimate the number of mils in deviation. This will give him the number of mils that the mortar needs to traverse.

16. The unit leader will add or subtract (LARS Rule-Left Add, Right Subtract) it to the mortars deflection in which it was mounted with. He would then announce his new deflection to the gunner. He will also determine any range corrections and data corrections.

EXAMPLE: Initial deflection 3200, unit leader's correction: L40 (LARS Rule) 3240

17.T he gunner will place the new deflection and elevation on the sight. Once he indexes the deflection and elevation, he will then sight back on the "T" bar, by traversing half the distance to the "T" bar and cross-leveling. The unit will continue steps 5 through 17 until the desired target kill is acquired.

SUPPORTING INDIVIDUAL TASKS

Task Number	Task Title
061-283-6003	Adjust Indirect Fire
071-074-0006	Manipulate a 60-mm, 81-mm, or 120-mm Mortar for Traversing and/or Searching Fire
091-257-0002	Conduct Preventive Maintenance Checks and Services (PMCS)
071-090-0002	Lay a 120-mm Mortar for Deflection and Elevation
071-323-4102	Lay a 60-mm Mortar for Deflection and Elevation

SUPPORTING INDIVIDUAL TASKS

Task Number	Task Title
071-074-0004	Engage Targets with a 60-mm, 81-mm, or 120-mm Mortar Using Direct Lay
071-074-0036	Adjust Mortar Fire Using Direct Alignment
071-074-0037	Fire a Mortar
071-074-0042	Lay a Mortar Using Direct Alignment
071-090-0004	Prepare 120-mm Mortar Ammunition for Firing
071-321-3905	Prepare 81-mm Mortar Ammunition for Firing
071-323-4106	Prepare 60-mm Mortar Ammunition for Firing
091-91F-2015	Correct Malfunctions of 81-mm M252 Mortar
091-91F-2016	Correct Malfunctions of 120-mm M120 Series Mortar

SUPPORTING COLLECTIVE TASKS

Task Number	Task Title
07-2-1342	Conduct Tactical Movement (Platoon-Company)

TASK: React to an IED Attack While Maintaining Movement (05-3-D0017)

CONDITIONS: The element conducts a mounted military operation when an improvised explosive device (IED) detonates.

CUE: An IED detonates within casualty-producing radius on the patrol, resulting in varying degrees of battle damage to the vehicles, equipment, and personnel.

STANDARDS: The element reacts to the IED attack by performing 5/25 meter checks. They use the 5Cs (confirm, clear, cordon, check, and control) to suppress enemy fire, set up security, evacuate casualties, recover disabled vehicles, submit an explosive hazards spot report, and exit the area.

TASK STEPS AND PERFORMANCE MEASURES
Unit personnel take the following actions:
1. Report the IED attack to the patrol (any Soldier can do this using the 3Ds: distance, direction, and description).
2. Establish 360-degree local security by directing the element to focus outward from the attack site.
3. If necessary, direct the element to the rally point based upon mission, enemy, terrain and weather, troops and support available, time available, and civil considerations (METT-TC) factors.
4. Employ tactical combat casualty care measures.
5. Evacuate casualties.
6. Conduct consolidation and reorganization at the rally point.
7. Direct element members to report the status of liquid, ammunition, casualties, and equipment (LACE) report.

SUPPORTING PRODUCTS

Product ID	Product Name
FM 3-21.75	Warrior Ethos and Soldier Combat Skills
ATTP 3-21.9	SBCT Infantry Rifle Platoon and Squad

SUPPORTING INDIVIDUAL TASKS

Task Number	Task Title
071-030-0004	Engage Targets with an MK 19 Grenade Machine Gun
071-054-0004	Engage Targets with an M136 Launcher
071-325-4407	Employ Hand Grenades
071-311-2130	Engage Targets with an M203 Grenade Launcher
071-010-0006	Engage Targets with an M249 Machine Gun
071-025-0007	Engage Targets with an M240B Machine Gun

SUPPORTING INDIVIDUAL TASKS

Task Number	Task Title
071-100-0030	Engage Targets with an M16-Series Rifle/M4-Series Carbine
071-326-0502	Move Under Direct Fire
071-326-5611	Conduct the Maneuver of a Squad
071-326-5630	Conduct Movement Techniques by a Platoon
071-121-4080	Send a Spot Report (SPOTREP)
061-283-1011	Engage Targets with Indirect Fires
113-571-1022	Perform Voice Communications

SUPPORTED COLLECTIVE TASKS

Task Number	Task Title
07-2-1090	Conduct a Movement to Contact (Platoon-Company)
07-2-1450	Secure Routes (Platoon-Company)
07-2-9002	Conduct a Bypass (Platoon-Company)
07-2-9009	Conduct a Withdrawal (Platoon-Company)

TASK: Perform Direct Lay of a Mortar (07-4-D9272)

CONDITIONS: The unit is operating where enemy contact is expected. The squad is ordered to engage a visible target by direct lay. The squad leader commands: "SQUAD, DIRECT LAY."

CUE: The squad leader initiates drill by giving the order: "SQUAD, DIRECT LAY."

STANDARDS: The squad is ordered to fire a target that they have direct line of sight with; they do not have FDC at this time. The squad leader commands: "UNIT DIRECT LAY (target) (direction) (distance)." The mortar is laid according to the initial fire command. The mortar is within two turns of center of traverse. The mortar is cross-leveled and all bubbles are centered. The vertical crossline on the sight unit is laid on the center of mass of the target.

Gunner levels the mortar, vertical crossline is laid center mass of target. Gunner announces: "UP."

TASK STEPS AND PERFORMANCE MEASURES

1. Unit leader commands: "UNIT, DIRECT LAY. (target) (direction) (distance)."
2. Gunner orients the mortar in the general direction of fire, which is identified by the unit leader.
3. Gunner places the deflection of 3,200 mils on the sight.
4. Unit leader determines and announces the shell and fuze, method of fire, charge, and elevation--for example, "HE quick, one round, charge three, elevation one one nine nine."
5. Gunner places the announced elevation on the sight.
6. Ammunition bearer prepares ammunition according to the announced shell and fuze, method of fire, and charge.
7. Gunner traverses the mortar, if necessary, within two turns of center of traverse so that the vertical crossline is laid on the center of the target .
8. Gunner levels the mortar so that all bubbles are centered and the vertical crossline is laid on the center of mass of the target.
9. Gunner announces, "UP."

NOTE: The gunner may use the turn method on the elevation crank when performing this drill.

SUPPORTING PRODUCTS

Product ID	Product Name
ATTP 3-21.90	Tactical Employment of Mortars

SUPPORTING INDIVIDUAL TASKS

Task Number	Task Title
061-283-6003	Adjust Indirect Fire
301-348-1050	Report Information of Potential Intelligence Value

SUPPORTED COLLECTIVE TASKS

Task Number	Task Title
07-3-3072	Fire a Mortar Fire for Effect Mission (Section-Unit)
07-3-3054	Fire a Mortar Adjust Fire Mission (Section-Unit)
07-3-5090	Process a Mortar Call for Fire Mission (Section-Unit)

TASK: Perform Hasty Lay of a 120-mm Mortar for Hipshoot (07-4-D9314)

ATTP 3-21.90   FM 3-22.90

CONDITIONS: The unit is moving tactically operating as part of a larger force and receives a request for suppressive fire. The unit leader commands, "FIRE MISSION."

CUE: The unit leader initiates drill by giving the order: "UNIT HASTY LAY."

STANDARDS: Unit leader halts the unit, commands "FIRE MISSION, and issues direction of fire (DOF)." Driver and assistant gunner exit the vehicle, position the trailer and remove the mortar from the trailer. Unit leader positions himself behind the mortar and takes an azimuth for direction of fire, the gunner and assistant gunner move the mortar left or right to bring the center of axis in alignment with the direction of fire.

1.   Mortar meets the following mounting and safety requirements:
    a.   Mask and overhead clearance are sufficient.
    b.   The front end of the baseplate is pointing in the direction of fire, and the barrel is locked to the baseplate with white line on top.
    c.   All bubbles on sight are leveled.
    d.   The clamp handle assembly is locked.
    e.   The baseplate is positioned correctly in relation to the unit leaders' directions.
    f.   The bipod legs are fully extended and the spread chain is taut.
    g.   The cross-level locking knob is hand tight.
2.   The mortar is laid on the mounting azimuth within 25 mils of the mounting azimuth. The traverse is within four turns from center-of-traverse. The traversing extension is locked in the center position. The sight is correctly set on the proper referred deflection and an elevation of 1,100 mils. The vertical crossline of the sight is within +/- 2 mils of the left edge of the near aiming post. The mortar unit is laid within the following time standards:
3.   Once the correct sight picture is attained the gunner announce his gun number "UP".

Time Standard: Day   Night
                              2 minutes   3 minutes

4.　If equipped with Mortar Fire Control Systems (MFCS) the time standards are as follows:

Section Laid Ready to Fire: 　　　　<u>Day</u>　　　　　<u>Night</u>
　　　　　　　　　　　　　　　　60 seconds　　1 minute 30 seconds

TASK STEPS AND PERFORMANCE MEASURES

1.　Unit leader determines DOF.
2.　Unit leader selects mortar position and halts the unit.
3.　Unit leader dismounts the vehicle and commands: "FIRE MISSION," then indicates DOF.
4.　Driver and ammunition bearers position trailer.
　　a.　The driver/ammunition bearer (first ammunition bearer) exits the vehicle and moves to the driver's side of the mortar trailer hitch.
　　b.　The second ammunition bearer exits the vehicle and moves to the passenger's side of the mortar trailer hitch.
　　c.　Together they unhook the trailer from the vehicle and position the trailer at the firing position with the baseplate toward the direction of fire.
5.　Assistant gunner and ammunition bearers remove mortar from trailer. (071-090-0004)
　　a.　The first ammunition bearer removes the muzzle plug.
　　b.　The assistant gunner places the aiming posts to the right side of the mortar.
　　c.　Once in position, both ammunition bearers raise the trailer until the baseplate is resting on the ground.
　　d.　The second ammunition bearer then releases the trailer while the first ammunition bearer continues to hold the trailer in place.
　　e.　The gunner places the sight to the left of the mortar.
　　f.　The gunner then moves up to the left side of the mortar and removes the lock release lever pin, and then unhooks the bipod chain from the eye on the bipod leg and drops it free.
　　g.　The gunner loosens the cross-level locking knob.
　　h.　The assistant gunner assumes his position on the right side of the mortar.
　　i.　The assistant gunner releases the lock release lever and the clamping catch, and swings the trailer bridge assembly out of the way.
　　j. The assistant gunner raises the bipod legs, and rotates them.
　　k.　The assistant gunner spreads the bipod legs until they are fully extended and the spread chain is taut.
　　l.. With the gunner standing on the left and second ammunition bearer on the right, together they both grasp the traversing mechanism.

TASK STEPS AND PERFORMANCE MEASURES

m.   With the assistant gunner holding the bipod legs just above the spread chain, They pull and guide the barrel forward away from the trailer.

n.   The assistant gunner guides the bipod legs to a point (about two feet) in front of the baseplate and lowers them to the ground and retightens the cross-level locking knob. At this point, the mortar is free of the trailer.

o.   The ammunition bearers move the trailer to the point selected by the unit leader away from the mortar position.

6.   Gunner and assistant gunner place mortar in firing configuration.

a.   Gunner straddles the barrel and loosens the clamp handle assembly.

b.   Gunner grasps under the recoil buffer housing assembly and pulls down, sliding the buffer housing assembly down the barrel until it rest against the lower collar stop.

c.   Gunner retightens the clamp handle assembly until it "Clicks," ensuring that the white line on the buffer housing assembly is aligned with the line on the barrel.

d.   Gunner places the safety mechanism to (F) fire.

e.   Assistant gunner checks for slack in the spread chain.

f. Assistant gunner ensures the cross-level locking knob is hand tight.

g.   Assistant gunner ensures the traversing mechanism is within four turns of center of traverse and the traversing extension is locked in the center position.

h.   The gunner removes the sight from the sight box. The gunner indexes a deflection of 3,200 mils and an elevation of 1,100 mils on the sight and places it in the dovetail slot of the bipod.

i. The unit leader positions himself behind the mortar with the compass aimed in the direction of fire. The assistant gunner moves the gun left or right, as ordered by the unit leader. The assistant gunner lifts the bipod and is guided by the unit leader. The gunner then levels the mortar for elevation and deflection. When the gunner is satisfied with the lay of the mortar, he announces, "UP."

7.   Unit loads and fires one round. (071-074-0037)

8.   The unit leader determines the referred deflection and gives it to the unit. (Normally, 2800 mils.) For example, REFER 2800, PLACE OUT AIMING POSTS. (071-090-0002)

9.   Gunner turns the deflection micrometer knob only and indexes deflection of 2,800 mils on his sight without disturbing the lay of the mortar.

TASK STEPS AND PERFORMANCE MEASURES

NOTE: Since this is an emergency mission (Hasty lay), only one aiming post is required to execute the mission. It will be placed out at 25 meters (approximate). After the mission is completed and it has been determined that the mortar will stay in its current position, then the unit will improve the mortar position. Improvements should include placing out both aiming post at 50 and 100 meters.

10. Once the correct sight picture is obtained, the gunner announces, "NUMBER (Number of gun) GUN UP."

SUPPORTING INDIVIDUAL TASKS

Task Number	Task Title
061-283-6003	Adjust Indirect Fire
091-257-0002	Conduct Preventive Maintenance Checks and Services (PMCS)
071-090-0002	Lay a 120-mm Mortar for Deflection and Elevation
071-074-0036	Adjust Mortar Fire Using Direct Alignment
071-074-0037	Fire a Mortar
071-074-0042	Lay a Mortar Using Direct Alignment
071-090-0004	Prepare 120-mm Mortar Ammunition for Firing
091-91F-2016	Correct Malfunctions of 120-mm M120-Series Mortar

SUPPORTING COLLECTIVE TASKS

Task Number	Task Title
07-3-2045	Reconnoiter a Mortar Firing Position (Section-Platoon)

TASK: Fire the Mortar (07-4-D9339)

CONDITIONS: The squad is in a firing position and receives the squad leader command: "HANG IT, and, FIRE." The squad leader initiates drill by giving the command: "HANG IT, and, FIRE."

CUE: The squad leader initiates drill by giving the command: HANG IT, and, FIRE

STANDARDS: The ammunition bearer selects the proper round. The round is inspected for dirt and serviceability. The propelling charge/increments are adjusted (removed or replaced) correctly. The fuze is set correctly. The round is inspected by the squad leader and assistant gunner before the squad leader gives the fire direction center (FDC) the command: "GUN UP." The round is loaded correctly by the assistant gunner. The assistant gunner releases the round correctly. The squad takes correct mortar firing positions. All commands are repeated.

TASK STEPS AND PERFORMANCE MEASURES

1.   Unit leaders record and issue the fire command to the squad.
2.   Unit repeats the fire commands.
3.   Ammunition bearers prepare the round(s) according to the fire command.
4.   Ammunition bearers prepare the round so the squad leader can inspect it before passing the round to the assistant gunner. Holding the round with both hands (palms up) near each end of the round body (not on the fuze or the charges).
NOTE: The ammunition bearer for a 60-mm mortar performs all functions of the assistant gunner.
5.   The ammunition bearer should hold the round with the fuze pointed to his left. The assistant gunner accepts the round from the ammunition bearer, with the right hand under the round and the left hand on top of the round, by pivoting his body to the left.
6.   The assistant gunner checks the round for correct charges, fuze tightness, and fuze setting.
7.   Once the assistant gunner has the round he keeps two hands on the round until it is fired.
8.   When both the gun and the round(s) have been determined safe and ready to fire, the squad leader gives the following command to the FDC, NUMBER (# of mortars) GUN UP.
9.   The assistant gunner is the only member of the mortar squad who loads and fires rounds.
10. The squad leader commands, HANG IT, and, FIRE, according to the method of fire given by the FDC.

TASK STEPS AND PERFORMANCE MEASURES

```
┌─────────────────────────────────────────────────────────┐
│                                                           │
│                       WARNING                             │
│                                                           │
│   The blast attenuator (BAD) must be installed on the     │
│   120-mm mortar before firing from the carrier.           │
│                                                           │
└─────────────────────────────────────────────────────────┘
```

11. Loading is done by holding the round out in front of the muzzle at the same angle as the cannon.

12. At the command, HANG IT, assistant gunners guide the round into the barrel (tail end first) to a point beyond the narrow portion of the body (about three-quarters of the round) ensuring not to hit the primer, charges, or disturbing the lay of the mortar.

13. Once the round is inserted into the barrel at the proper distance, assistant gunners shout back: "NUMBER (number of mortar) GUN, HANGING."

14. At the command, FIRE, assistant gunners release the round by pulling both their hands down alongside and away from the outside of the barrel. Assistant gunners must ensure they do not take their hands across the muzzle of the cannon as they drop the round.

15. Once the round is released by the assistant gunner, the gunner and assistant gunner each position themselves according to the safety requirements for firing outlined in the appropriate TM.

16. Assistant gunners pivot to their left and down toward the ammunition bearer to accept the next round to be fired.

17. Subsequent rounds are fired based on the FDC fire commands.

18. Assistant gunners ensure the round has fired safely before they attempt to load the next round.

19. Assistant gunners and gunners, as well as the remainder of the mortar crew, must keep their upper body below the muzzle until the round fires to avoid muzzle blast.

20. During a fire for effect (FFE), the gunner attempts to level all bubbles between each round ensuring his upper body is away from the mortar and below the muzzle when the assistant gunner announces: "HANGING," for each round fired.

21. Assistant gunners inform the squad leader when all rounds for the fire mission are expended and the squad leader informs the FDC when all of the rounds were completed. For example, NUMBER TWO GUN, ALL ROUNDS COMPLETE.

SUPPORTING PRODUCTS

Product ID	Product Name
ATTP 3-21.90	Tactical Employment of Mortars

SUPPORTING INDIVIDUAL TASKS

Task Number	Task Title
061-283-6003	Adjust Indirect Fire
301-348-1050	Report Information of Potential Intelligence Value

SUPPORTED COLLECTIVE TASKS

Task Number	Task Title
07-3-3072	Fire a Mortar Fire for Effect Mission (Section-Unit)
07-3-3054	Fire a Mortar Adjust Fire Mission (Section-Unit)
07-3-5090	Process a Mortar Call for Fire Mission (Section-Unit)

TASK: React to Indirect Fire (07-3-D9504)

CONDITIONS:

Dismounted. The unit moves, conducting operations. Any Soldier gives the alert, INCOMING, or a round impacts nearby.

Mounted. The platoon/section is stationary or moves, conducting operations. The alert, INCOMING, comes over the radio or intercom or rounds impact nearby.

CUE: This drill begins when any member alerts, INCOMING, or a round impacts.

STANDARDS:

Dismounted. Soldiers immediately seek the best available cover. The unit moves out of area to the designated rally point after the impacts.

Mounted. When moving, drivers immediately move their vehicles out of the impact area in the direction and distance ordered. If stationary, drivers start their vehicles and move in the direction and distance ordered. Unit leaders report the contact to higher headquarters (HQ).

TASK STEPS AND PERFORMANCE MEASURES

1. **Dismounted.** Unit personnel take the following actions:
    a. Any Soldier announces, INCOMING!
    b. Soldiers immediately assume the prone position or move to immediate available cover during initial impacts.
    c. The unit leader orders the unit to move to a rally point by giving a direction and distance.
    d. After the impacts, Soldiers move rapidly in the direction and distance to the designated rally point.
    e. The unit leaders report the contact to higher HQ.
2. **Mounted.** Unit personnel take the following actions:
    a. Any Soldier announces, INCOMING!
    b. Vehicle commanders repeat the alert over the radio.
    c. The leaders give the direction and link-up location over the radio.
    d. Soldiers close all hatches if applicable to the vehicle type; gunners stay below turret shields or get down into vehicle.
    e. Drivers move rapidly out of the impact area in the direction ordered by the leader.
    f. Unit leaders report the contact to higher HQ.

SUPPORTING PRODUCTS

Product ID	Product Name
FM 3-21.8	The Infantry Rifle Platoon and Squad

SUPPORTING PRODUCTS

Product ID   **Product Name**
FM 3-21.75   Warrior Ethos and Soldier Combat Skills
ATTP 3-21.9   SBCT Infantry Rifle Platoon and Squad

SUPPORTING INDIVIDUAL TASKS

Task Number   **Task Title**
071-326-3002   React to Indirect Fire While Mounted
113-571-1022   Perform Voice Communications

SUPPORTED COLLECTIVE TASKS

Task Number   **Task Title**
07-2-3000   Conduct Support by Fire (Platoon-Company)
07-2-9004   Conduct a Delay (Platoon-Company)
07-2-9009   Conduct a Withdrawal (Platoon-Company)
17-2-9225   Conduct a Screen (Platoon-Company)

TASK: Lay 81mm Mortar for Large Deflection and Elevation Change (07-4-D9275)

CONDITIONS: The unit receives a fire command requiring a deflection change of more than 200 mils but less than 300 mils and an elevation change of more than 100 mils but less than 200 mils. The unit leader gives the command for the new deflection and elevation change.

CUE: This drill begins on the command of the squad leader.

STANDARDS: Correct deflection and elevation are emplaced on the sight. Bubbles are centered on the sight. The vertical crossline is within +/- 2 mils of the left edge of the aiming post (or of a compensated sight picture). The traversing bar must be within two turns of left and right of center.

1.   The mortar meets the following mounting and safety requirements.
  a. The mask and overhead clearance are sufficient.
  b. The open end of the socket cap is pointing in the direction of fire, and the barrel is locked to the baseplate.
  c. (60mm Specific). The coarse cross-level nut on the locking sleeve is tightened.
  d. (60mm Specific). The collar assembly is positioned on the correct saddle of the barrel, determined by the announced elevation.
  e. (60mm Specific). The collar is secured onto the barrel by the collar knob.
  f. All bubbles on sight are level.
  g. The sight has correct setting for deflection and elevation.
  h. (60mm Specific). The traversing bearing is no more than two turns left or right of center.
  i. The baseplate is positioned correctly in relation to the baseplate stake.
  j. (60mm Specific). Bipod legs are fully extended and the spread cable is taut.
2.   The gunner announces, UP, within 60 seconds following the last digit of the elevation element.

TASK STEPS AND PERFORMANCE MEASURES:
1.   Unit leader issues fire command to the squad.
2.   Unit members repeat fire command to the squad leader.
3.   Gunner places the announced deflection and elevation on the sight.
4.   Gunner turns the elevation hand crank until the elevation bubble is approximately centered. He centers the buffer carrier.
5.   Assistant gunner moves into position to the front of the mount, and kneels.

TASK STEPS AND PERFORMANCE MEASURES:

6.   Assistant gunner grasps the mount (palms up) and lifts legs until they are clear of the ground to permit lateral movement.

7.   Gunner and assistant gunner move the mortar and align the sight on the aiming posts.

8.   Assistant gunner tries to maintain the traversing gear assembly on a horizontal plane. To make the shift, the gunner places the fingers of the right hand in the muzzle and his left hand on the elevating leg assembly.

9.   Gunner announces to the assistant gunner to lower the mount when the vertical crossline is within 20 mils of the aiming post.

10. Assistant gunner centers the elevation bubble and cross-levels the weapon.

11. If the gunner finds he does not have a proper sight picture, he performs the following tasks:

   a. Traverses half the distance to get a proper sight picture.

   b. Levels the sight.

   c. Divides the distance in half again and levels again.

   d. Repeats the process until he has a proper sight picture.

12. Gunner announces, "UP."

SUPPORTING INDIVIDUAL TASKS

Task Number	Task Title
061-283-6003	Adjust Indirect Fire
071-074-0006	Manipulate a 60-mm, 81-mm, or 120-mm Mortar for Traversing and/or Searching Fire
091-257-0002	Conduct Preventive Maintenance Checks and Services (PMCS)
071-090-0002	Lay a 120-mm Mortar for Deflection and Elevation
071-323-4102	Lay a 60-mm Mortar for Deflection and Elevation

SUPPORTED COLLECTIVE TASKS

Task Number	Task Title
07-3-3054	Fire a Mortar Adjust Fire Mission (Section-Platoon)

TASK: Conduct the 5Cs (05-3-D0016)

CONDITIONS: The element conducts a mounted or dismounted military patrol when an improvised explosive device (IED) is identified or detonates.

CUE: This is done when a possible or suspected IED is identified, an explosive device is detonated, or while conducting a security halt (mounted or dismounted).

STANDARDS: The element conducts the 5Cs (confirm, clear, cordon, check, control) correctly, ensuring the area is clear of any nonessential personnel, secondary or tertiary IEDs have been confirmed and identified, a cordon has been established, and personnel access to the area is under control.

TASK STEPS AND PERFORMANCE MEASURES

NOTE: Conduct the 5C's; these are not order specific and can be done concurrently.

1. Confirms there is a requirement for explosive ordnance disposal (EOD) when encountering a suspected or known IED.
2. Clears all personnel from the area to a tactically safe position and distance from the potential IED.
3. Cordons the area.
4. Checks the immediate area for secondary/tertiary devices around the incident control point (ICP) and cordon using the 5/25 meter checks.
5. Controls the area inside the cordon to ensure only authorized access.

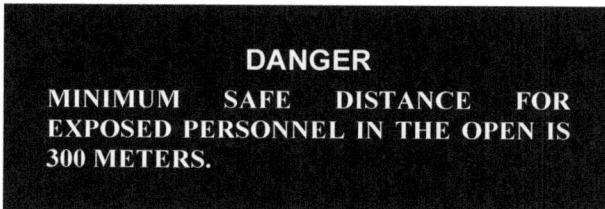

> **DANGER**
>
> MINIMUM SAFE DISTANCE FOR EXPOSED PERSONNEL IN THE OPEN IS 300 METERS.

SUPPORTING PRODUCTS

Product ID	Product Name
FM 3-21.75	Warrior Ethos and Soldier Combat Skills

SUPPORTED INDIVIDUAL TASKS

Task Number	Task Title
081-831-0101	Request Medical Evacuation

SUPPORTED INDIVIDUAL TASKS

Task Number	Task Title
081-831-1003	Perform First Aid to Clear an Object Stuck in the Throat of a Conscious Casualty
081-831-1005	Perform First Aid to Prevent or Control Shock
081-831-1007	Perform First Aid for Burns
081-831-1025	Perform First Aid for an Open Abdominal Wound
081-831-1026	Perform First Aid for an Open Chest Wound
081-831-1032	Perform First Aid for Bleeding and/or Severed Extremity
081-831-1033	Perform First Aid for an Open Head Wound
081-831-1034	Perform First Aid for a Suspected Fracture
081-831-1046	Transport a Casualty
113-571-1022	Perform Voice Communications
805C-PAD-2060	Report Casualties

SUPPORTED COLLECTIVE TASKS

Task Number	Task Title
07-2-5027	Conduct Consolidation and Reorganization (Platoon-Company)
08-2-0003	Treat Casualties
08-2-0004	Evacuate Casualties

TASK: Evacuate a Casualty (Dismounted and Mounted) (07-3-D9507)

CONDITIONS: The unit is stationary or moves, conducting operations. A Soldier has been injured and must be evacuated. All enemy in the area have been suppressed, neutralized, or destroyed, and local security is established. Some iterations of this drill should be performed in mission-oriented protective posture 4 (MOPP 4).

CUE: This drill begins when a unit member is injured and must be evacuated or the leader directs his personnel to conduct the drill.

STANDARDS: Element members conduct first aid and evacuate the casualties without dropping or causing further injury to the casualties. If necessary, the unit leader, combat medic, or any Soldier requests medical evacuation (MEDEVAC) and reports the contact to higher headquarters (HQ).

TASK STEPS AND PERFORMANCE MEASURES

1. Element members conduct first aid and evacuate the casualties without dropping or causing further injury to the casualties.
2. Drill is conducted while dismounted. Unit personnel take the following actions:

    a. Any unit member provides initial first aid (self-aid/buddy aid).

    b. Any unit combat lifesaver provides enhanced first aid or combat medic provides emergency medical treatment if necessary.

    c. The unit leader, combat medic, or any Soldier requests MEDEVAC using the 9-Line MEDEVAC request if necessary.

    d. The unit aid and litter team or designated members evacuate casualties to the casualty collection point (CCP) or patient collecting point (PCP) and request MEDEVAC. They take the following actions:

        (1) Remove key operational items and equipment (maps, simple key loader [SKL]/automated network control devices [ANCD], position-locating devices, laser pointers, and other sensitive items).

        (2) Account for the weapons and ammunition of casualties according to the unit standing operation procedures (SOPs).

        (3) Complete DD Form 1380, *U.S. Field Medical Card*, and unit leaders or any member complete Department of the Army (DA) Form 1156, *Casualty Feeder Card*.

        (4) Evacuate casualty to the CCP, PCP, or aid station using litters, one or two man carry, or by having casualties with minor wounds walk.

3. Drill is conducted while mounted. Unit personnel take the following actions:

    a. Crew/occupants provide initial first aid (self-aid/buddy aid).

TASK STEPS AND PERFORMANCE MEASURES

b. Any unit combat lifesaver, combat medic, or designated Soldier moves to the vehicle to provide first aid or enhanced first aid (self-aid, buddy aid, and combat lifesaver) and emergency medical treatment (EMT) (combat medic) and then evacuates the casualty.

c. Designated Soldiers remove the casualty from the vehicle so as not to cause further injury. They take the following actions:

(1) Remove all key operational items and equipment (maps, simple key loader [SKL]/automated network control devices [ANCD], position-locating devices, and all other sensitive items).

(2) Account for the weapons and ammunition of casualties according to unit SOPs.

(3) Complete DD Form 1380, and DA Form 1156.

(4) Evacuate casualties to the CCP or PCP and request MEDEVAC (9-line MEDEVAC request) or evacuate directly to the aid station using available vehicle assets.

4. Unit leaders report the contact and casualties according to unit SOPs to higher HQ.

SUPPORTING PRODUCTS

Product ID	Product Name
FM 3-21.8	The Infantry Rifle Platoon and Squad
FM 3-21.75	Warrior Ethos and Soldier Combat Skills

SUPPORTED INDIVIDUAL TASKS

Task Number	Task Title
081-831-0101	Request Medical Evacuation
081-831-1003	Perform First Aid to Clear an Object Stuck in the Throat of a Conscious Casualty
081-831-1005	Perform First Aid to Prevent or Control Shock
081-831-1007	Perform First Aid for Burns
081-831-1025	Perform First Aid for an Open Abdominal Wound
081-831-1026	Perform First Aid for an Open Chest Wound
081-831-1032	Perform First Aid for Bleeding and/or Severed Extremity
081-831-1033	Perform First Aid for an Open Head Wound
081-831-1034	Perform First Aid for a Suspected Fracture
081-831-1046	Transport a Casualty
113-571-1022	Perform Voice Communications
805C-PAD-2060	Report Casualties

SUPPORTED COLLECTIVE TASKS

Task Number	Task Title
07-2-5027	Conduct Consolidation and Reorganization (Platoon-Company)
08-2-0003	Treat Casualties
08-2-0004	Evacuate Casualties

TASK: React to Air Attack Drill (17-3-D8004)

CONDITIONS: While operating in a tactical environment, the platoon or section identifies threat aircraft, requiring it to take either passive or active air defense measures. The platoon is digitally connected (if equipped) with higher headquarters (HQ) via Force XXI Battle Command Brigade and Below (FBCB2). Some iterations of this task should be performed in mission-oriented protective procedures 4 (MOPP 4).

CUE: Any Soldier gives an oral or visual signal for a chemical attack or when a chemical alarm activates.

STANDARDS: The platoon or section executes appropriate air defense measures and prevents the aircraft from effectively engaging and/or observing the platoon/section. The platoon reports to higher HQ. No friendly unit suffers casualties or equipment damage as a result of fratricide.

TASK STEPS AND PERFORMANCE MEASURES
1. The vehicle or individual who identifies threat aircraft alerts the platoon with a contact report containing these elements:
   a. Contact.
   b. Bandit(s).
   c. Cardinal direction (specify: north, south, east, or west).
2. Platoon/section leaders analyze situation to determine whether the platoon is in the direct path of and/or is the target of the threat aircraft. They take the following actions:
   a. Order passive air defense measures when the platoon/section is not in the path of or target of the threat aircraft.
OR
   b. Order active air defense measures when the platoon is in the path of or target of the threat aircraft.
3. Platoons or sections execute passive air defense measures as necessary. They take the following actions:
   a. On order of platoon/section leaders, move to covered and concealed positions, maintaining a minimum of 100 meters between vehicles and halts.
   b. Prepare to engage on order of platoon/section leader.
   c. Scan for follow-on aircraft.
NOTE: Higher HQ may order the platoon or section to continue movement.
4. Platoons execute active air defense measures as necessary. They take the following actions:
   a. If in the direct path of flight, move away from the path of flight as fast as possible, moving at a 45-degree angle toward the attacking aircraft.

TASK STEPS AND PERFORMANCE MEASURES

b. Maintain at least 100-meter intervals and avoid creating a linear target for the attacking aircraft.

c. Orient on the aim point designated by the platoon/section leader and engage the aircraft with a high volume of machine gun fire using the proper lead technique for the type of aircraft and direction of movement.

d. Move quickly to covered and concealed positions and halts.

e. Remain in covered and concealed positions, as required.

f. Scan for follow-on aircraft.

5. Platoon leaders/platoon sergeants (PSGs) report situation to higher HQ as necessary. They send:

a. Spot report (SPOTREP).

b. (D) Updated situation reports (SITREP), as necessary.

SUPPORTING PRODUCTS

Product ID	Product Name
FM 3-20.98	Reconnaissance and Scout Platoon
FM 3-20.971	Reconnaissance and Cavalry Troop
FM 3-20.15	Tank Platoon

SUPPORTING INDIVIDUAL TASKS

Task Number	Task Title
171-121-4017	Supervise Tank Platoon Formations and Drills
171-121-4051	Prepare a Situation Report (SITREP)
171-121-4057	Perform Techniques of Movement

SUPPORTING COLLECTIVE TASKS

Task Number	Task Title
07-2-3000	Conduct Support by Fire (Platoon-Company)
07-2-9001	Conduct an Attack (Platoon-Company)
07-2-9003	Conduct a Defense (Platoon-Company)
07-2-9012	Conduct a Relief in Place (Platoon-Company)
17-5-5585	Engage Multiple Machine Gun Targets on a M1-Series Tank
17-5-5590	Conduct Main Gun Misfire Procedures on a M1-Series Tank
17-5-5622	Engage Targets with the Main Gun from a M1-Series Tank
17-5-8006	React to an Antitank Guided Missile (ATGM)

TASK: React to Ambush (Near) (07-3-D9502)

CONDITIONS: (Dismounted/Mounted) - The unit is moving tactically, conducting operations. The enemy initiates contact with direct fire within hand grenade range. All or part of the unit is receiving accurate enemy direct fire.

CUE: This drill begins when the enemy initiates ambush within hand grenade range.

STANDARDS:

Dismounted. Soldiers in the kill zone immediately return fire on known or suspected enemy positions and assault through the kill zone. Soldiers not in the kill zone locate and place "well-aimed" suppressive fire on the enemy. The unit assaults through the kill zone and destroys the enemy.

Mounted. Vehicle gunners immediately return fire on known or suspected enemy positions as the unit continues to move out of the kill zone. Soldiers on disabled vehicles in the kill zone dismount, occupy covered positions and engage the enemy with accurate fire. Vehicle gunners and Soldiers outside the kill zone suppress the enemy. The unit assaults through the kill zone and destroys the enemy. The unit leader reports the contact to higher headquarters (HQ).

TASK STEPS AND PERFORMANCE MEASURES
1.   Dismounted (See Figure 1.); takes the following actions:

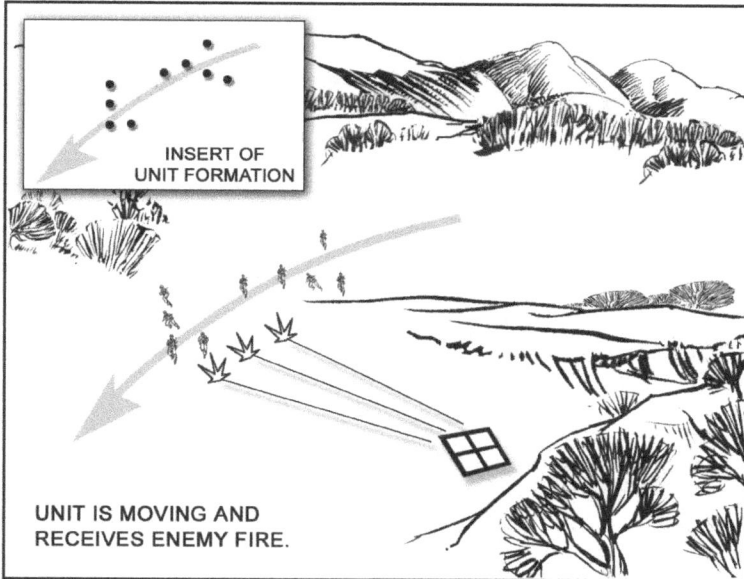

Figure 1. React to ambush (near) (dismounted)

a. Soldiers in the kill zone execute one of the following two actions:

(1) Return fire immediately. If cover is not available, immediately, without order or signal, assault through the kill zone.

(2) Return fire immediately. If cover is available, without order or signal, occupy the nearest covered position, and throw smoke grenades. (See Figure 2.)

TASK STEPS AND PERFORMANCE MEASURES

SOLDIERS IN THE KILL ZONE
IMMEDIATELY RETURN FIRE.

Figure 2. React to ambush (near) (dismounted) (continued)

b. Soldiers in the kill zone assault through the ambush using fire and movement.

c. Soldiers not in the kill zone identify the enemy location, place "well-aimed" suppressive fire on the enemy's position and shift fire as Soldiers assault the objective.

d. Soldiers assault through and destroy the enemy position. (See Figure 3.)

TASK STEPS AND PERFORMANCE MEASURES

SOLDIERS IN SUPPORT POSITION
SHIFT FIRES AS SOLDIERS IN KILL
ZONE ASSAULT ENEMY POSITION(S).

Figure 3. React to ambush (near) (dismounted) (continued)

e. The unit leader reports the contact to higher HQ.

2. Mounted; takes the following actions:

a. Vehicle gunners in the kill zone immediately return fire and deploy vehicle smoke, while moving out of the kill zone.

b. Soldiers in disabled vehicles in the kill zone immediately obscure themselves from the enemy with smoke, dismount if possible, seek covered positions, and return fire.

c. Vehicle gunners and Soldiers outside of the kill zone identify the enemy positions, place "well-aimed" suppressive fire on the enemy, and shift fire as Soldiers assault the objective.

d. The unit leader calls for and adjusts indirect fire and request close air support according to METT-TC.

e. Soldiers in the kill zone assault through the ambush and destroy the enemy.

f. The unit leader reports the contact to higher HQ.

SUPPORTING PRODUCTS

Product ID	Product Name
FM 3-21.75	Warrior Ethos and Soldier Combat Skills
FM 3-21.8	The Infantry Rifle Platoon and Squad
ATTP 3-21.9	SBCT Infantry Rifle Platoon and Squad

SUPPORTING INDIVIDUAL TASKS

Task Number	Task Title
071-000-0006	React to Man-to-Man Contact
071-030-0004	Engage Targets with an MK 19 Grenade Machine Gun
071-054-0004	Engage Targets with an M136 Launcher
071-311-2130	Engage Targets with an M203 Grenade Launcher
071-325-4407	Employ Hand Grenades
071-705-0007	Engage Targets with an M16-Series Rifle/M4-Series Carbine using an M68 Reflex Sight (Close Combat Optic)
071-010-0006	Engage Targets with an M249 Machine Gun
061-283-6003	Adjust Indirect Fire
071-326-0502	Move Under Direct Fire
071-326-5606	Select an Overwatch Position
071-410-0002	React to Direct Fire While Mounted
113-571-1022	Perform Voice Communications
081-831-1001	Evaluate a Casualty (Tactical Combat Casualty Care)
081-831-1003	Perform First Aid to Clear an Object Stuck in the Throat of a Conscious Casualty
081-831-1005	Perform First Aid to Prevent or Control Shock
071-326-0608	Use Visual Signaling Techniques
071-326-0510	React to Indirect Fire While Dismounted
071-326-3002	React to Indirect Fire While Mounted
071-326-0501	Move as a Member of a Fire Team

SUPPORTING COLLECTIVE TASKS

Task Number	Task Title
07-2-1342	Conduct Tactical Movement (Platoon-Company)
07-2-1189	Conduct a Dismounted Tactical Road March (Platoon-Company)
07-2-1198	Conduct a Mounted Tactical Road March (Platoon-Company)

TASK: React to Ambush (Far) (07-3-D9503)

CONDITIONS:

Dismounted/mounted. The platoon/squad/section moves tactically, conducting operations. The enemy initiates contact with direct and indirect fire.

CUE: This drill begins when the enemy initiates ambush with direct and indirect fire.

STANDARDS:

Dismounted. The unit immediately returns fire and occupies covered and/or concealed positions. The unit moves out of the kill zone, locates the enemy position, and conducts fire and maneuver to destroy the enemy.

Mounted. Vehicle gunners immediately return fire on known or suspected enemy positions as the unit continues to move out of the kill zone. The unit leader reports the contact to higher headquarters (HQ).

TASK STEPS AND PERFORMANCE MEASURES

1. Dismounted. (See Figure 1.) Unit personnel take the following actions:

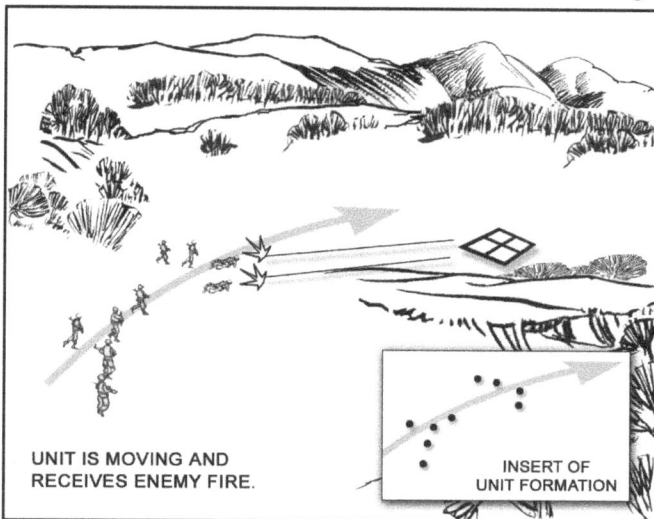

UNIT IS MOVING AND RECEIVES ENEMY FIRE.

INSERT OF UNIT FORMATION

Figure 1. React to ambush (far) (dismounted)

a. Soldiers receiving fire immediately return fire, seek cover, establish a support by fire, and suppress the enemy position(s).

b. Soldiers not receiving fire move along a covered and concealed route to the enemies flank to assault the enemy position. (See Figure 2.)

TASK STEPS AND PERFORMANCE MEASURES

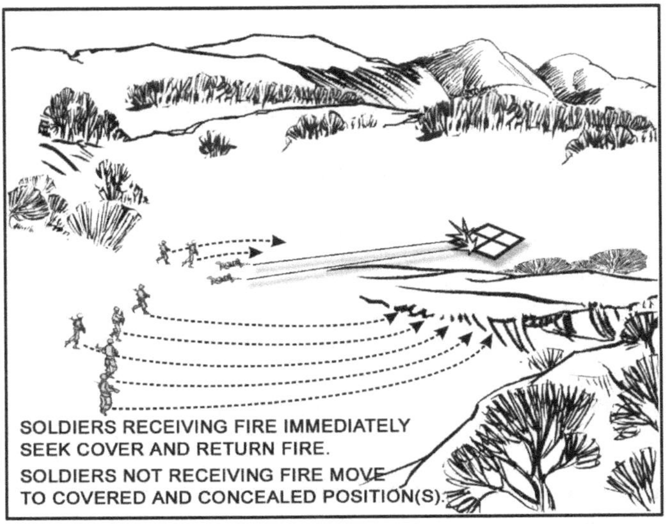

SOLDIERS RECEIVING FIRE IMMEDIATELY SEEK COVER AND RETURN FIRE.
SOLDIERS NOT RECEIVING FIRE MOVE TO COVERED AND CONCEALED POSITION(S).

Figure 2. React to ambush (far) (dismounted) (continued)

c. Unit leaders or forward observers call for and adjust indirect fires and close air support, if available. On order, the unit leaders or forward observers lift or shift fires to isolate the enemy position or to attack them with indirect fires as they retreat.

d. Soldiers in the kill zone shift suppressive fires as the assaulting Soldiers fight through and destroy the enemy. (See Figure 3.)

TASK STEPS AND PERFORMANCE MEASURES

SOLDIERS IN KILL ZONE SHIFT FIRE
AS ASSAULTING SOLDIERS DESTROY
ENEMY.

Figure 3. React to ambush (far) (dismounted) (continued)

    e. Unit leaders report the contact to higher HQ.

2.   Mounted. Unit personnel take the following actions:

    a. Gunners and personnel on vehicles immediately return fire.

    b. If the roadway is clear, they move all vehicles through the kill zone.

    c. Soldiers on the lead vehicle deploy vehicle smoke to obscure the enemy's view of the kill zone.

    d. The vehicle commander, in disabled vehicles, orders Soldiers to dismount according to the variables of mission, enemy, terrain and weather, troops and support available-time available and civil considerations (METT-TC) and sets up security while awaiting recovery.

    e. The remainder of the unit follows the lead vehicle out of the kill zone while continuing to suppress the enemy.

    f. Unit leaders report the contact to higher HQ.

SUPPORTING PRODUCTS

Product ID	Product Name
FM 3-21.8	The Infantry Rifle Platoon and Squad
FM 3-21.75	Warrior Ethos and Soldier Combat Skills
ATTP 3-21.9	SBCT Infantry Rifle Platoon and Squad

SUPPORTING INDIVIDUAL TASKS

Task Number	Task Title
071-326-0502	Move Under Direct Fire
071-054-0004	Engage Targets with an M136 Launcher
071-311-2130	Engage Targets with an M203 Grenade Launcher
071-325-4407	Employ Hand Grenades
113-571-1022	Perform Voice Communications
071-326-5606	Select an Over-watch Position
071-326-5611	Conduct the Maneuver of a Squad
071-410-0002	React to Direct Fire While Mounted
071-100-0003	Engage Targets with an M4 or M4A1 Carbine
071-010-0006	Engage Targets with an M249 Machine Gun
071-313-3454	Engage Targets with a Caliber .50 M2 Machine Gun
071-025-0007	Engage Targets with an M240B Machine Gun
081-831-1001	Evaluate a Casualty (Tactical Combat Casualty Care)
081-831-1003	Perform First Aid to Clear an Object Stuck in the Throat of a Conscious Casualty

SUPPORTED COLLECTIVE TASKS

Task Number	Task Title
19-3-2007	Conduct Convoy Security
07-2-1342	Conduct Tactical Movement (Platoon-Company)
07-2-1189	Conduct a Dismounted Tactical Road March (Platoon-Company)
07-2-1198	Conduct a Mounted Tactical Road March (Platoon-Company)
19-2-2004	Perform Convoy Security

TASK: Establish Security at the Halt (07-3-D9508)

CONDITIONS):

Dismounted/mounted. The unit moves tactically, conducting operations. An unforeseen event causes the unit to halt. Enemy contact is possible.

CUE: This drill begins when the unit must halt and enemy contact is possible, or the unit leader initiates drill by giving the order, HALT.

STANDARDS

Dismounted. Soldiers stop movement and clear the area per unit standing operating procedures (SOPs). (An example technique is the 5-25 meters; each Soldier immediately scans 5 meters around his position and then searches out to 25 meters based on the duration of the halt). Soldiers occupy covered and concealed positions, and maintain dispersion and all-round security.

Mounted. Vehicle commanders direct their vehicles into designated positions, using available cover and concealment. Soldiers dismount in the order specified and clear the area per unit SOPs. (An example technique is the 5-25 meters; each Soldier immediately scans 5 meters around his position and then searches out to 25 meters based on the duration of the halt). Platoon/section members maintain dispersion and all-round security.

TASK STEPS AND PERFORMANCE MEASURES

1. Dismounted. Unit personnel take the following actions:
    a. The unit leader gives the arm-and-hand signal to halt.
    b. Soldiers establish local security. They take the following actions:
        (1) Assume hasty fighting positions using available cover and concealment.
        (2) Inspect and clear their immediate area (Example: using the 5-25 technique).
        (3) Establish a sector of fire for their assigned weapon (Example: using 12 o'clock as the direction the Soldier is facing, the Soldier's sector of fire could be his 10 o'clock to 2 o'clock).
    c. Element leaders adjust positions as necessary. They take the following actions:
        (1) Inspect and clear their element area.
        (2) Ensure Soldiers sector of fire overlap.
        (3) Coordinate sectors with the elements on their left and right.
    d. Unit leaders report the situation to higher headquarters (HQ).
2. Mounted. Unit personnel take the following actions:

TASK STEPS AND PERFORMANCE MEASURES

a.   Unit leaders give the order over the radio to stop movement and establish security (See Figure 1.)

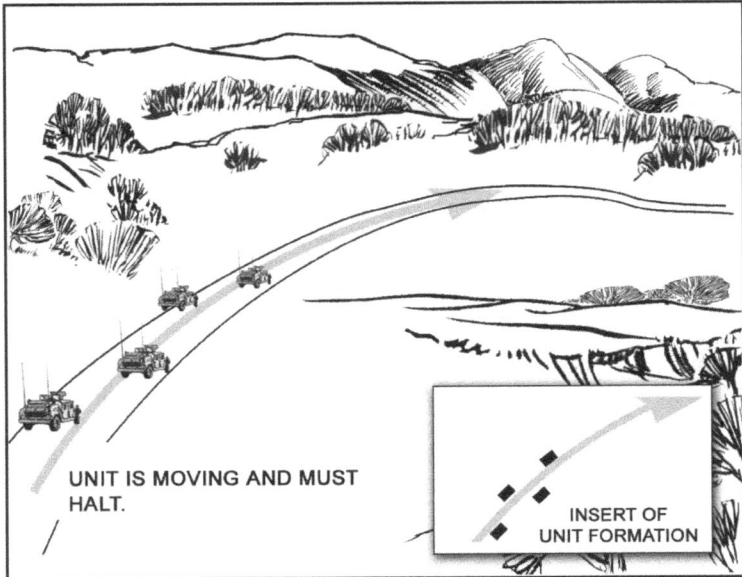

UNIT IS MOVING AND MUST HALT.

INSERT OF UNIT FORMATION

Figure 1. Establish security at the halt

b.   The unit halts in the herringbone or coil formation according to the unit SOPs. (See Figures 2 and 3.)

TASK STEPS AND PERFORMANCE MEASURES

VEHICLES MOVE INTO POSITIONS AND
SOLDIERS DISMOUNT TO PROVIDE
SECURITY.

**Figure 2. Establish security at the halt (mounted) (herringbone)
(continued)**

TASK STEPS AND PERFORMANCE MEASURES

VEHICLES MOVE INTO POSITIONS AND
SOLDIERS DISMOUNT TO PROVIDE
SECURITY.

**Figure 3. Establish security at the halt (mounted) (coil)
(continued)**

c.   Each vehicle commander ensures his vehicle is correctly positioned, using cover and concealment, and the crew served weapon is manned and scanning.

d.   Vehicle commanders order Soldiers to dismount to provide local security.

e.   Soldiers dismount and establish local security. They take the following actions:

(1) Move to a covered and concealed position as designated by the leader.

(2) Inspect and clear their immediate area (example: using the 5-25 technique).

(3) Establish a sector of fire for their assigned weapons.

f.   Dismount element leaders adjust positions as necessary.

g.   Unit leaders report the situation to higher HQ.

SUPPORTING PRODUCTS

Product ID	Product Name
FM 3-21.8	The Infantry Rifle Platoon and Squad
FM 3-21.75	Warrior Ethos and Soldier Combat Skills
ATTP 3-21.9	SBCT Infantry Rifle Platoon and Squad

SUPPORTING INDIVIDUAL TASKS

Task Number	Task Title
113-571-1022	Perform Voice Communications
551-88M-0005	Operate a Vehicle in a Convoy
071-326-0513	Select Temporary Fighting Positions
071-326-0608	Use Visual Signaling Techniques
071-331-0801	Challenge Persons Entering Your Area
071-331-0815	Practice Noise, Light, and Litter Discipline
071-331-1004	Perform Duty as a Guard
191-376-4114	Control Entry to and Exit From a Restricted Area
191-376-5140	Search a Vehicle for Explosive Devices or Prohibited Items at an Installation Access Control Point
191-376-5151	Control Access to a Military Installation
551-001-1040	Perform 5/25-Meter Scans
551-001-1041	Establish Security While Mounted (if applicable)
551-001-1042	Dismount a Vehicle
551-001-1043	React to Vehicle Rollover
551-88M-1658	Prepare Vehicle for Convoy Operations

SUPPORTED COLLECTIVE TASKS

Task Number	Task Title
19-3-2007	Conduct Convoy Security
07-2-1189	Conduct a Dismounted Tactical Road March (Platoon-Company)
07-2-1198	Conduct a Mounted Tactical Road March (Platoon-Company)
07-2-1342	Conduct Tactical Movement (Platoon-Company)
07-2-9011	Conduct Tactical Movement in an Urban Area (Platoon-Company)

TASK: Mount the Mortar Carrier (07-4-D9280.P)

CONDITIONS: The unit must displace to a new firing position. The unit leader commands, MOUNT. The unit leader initiates drill by giving the order to "Mount."

CUE: The unit leader commands, MOUNT.

STANDARDS: The mortar carriers move into covered and concealed positions. When the command, MOUNT, is given, the unit mounts the mortar carrier within one minute with all mission-essential equipment.

TASK STEPS AND PERFORMANCE MEASURES

1. Unit leader moves to the exact location where the mortar carrier will halt.
2. Unit leader announces, "Prepare to mount."
3. Driver moves to the mortar carrier and mounts.
4. Driver moves mortar carrier to the location as directed by the unit leader. (071-410-0018)
5. Driver halts the mortar carrier.
6. Unit leader commands, MOUNT.
7. Assistant gunner mounts and secures the barrel in position.
8. Gunner mounts and secures the mount in position.
9. Ammunition bearer secures the baseplate in position and mounts the mortar carrier.
10. Unit leader secures the sight case and aiming posts.
11. Unit leader ensures all personnel and equipment are accounted for and properly secured.
12. Unit leader mounts the mortar carrier.

SUPPORTING INDIVIDUAL TASKS

Task Number	Task Title
071-074-0006	Manipulate a 60-mm, 81-mm, or 120-mm Mortar for Traversing

SUPPORTING COLLECTIVE TASKS

Task Number	Task Title
07-3-1306	Displace a Mortar Unit by Echelon (Platoon/Squad)

TASK: React to Nuclear Attack (07-3-D9483)

CONDITIONS: The current enemy situation outlines a threat of tactical nuclear capabilities. With little-to-no warning, the squad/platoon encounters a brilliant flash of light while conducting tactical operations. Some portion of the unit moves in the assigned area while others are dismounted, either stationary or moving at the time of the attack.

CUE: The drill begins when any member sees a brilliant flash of light or a leader directs his personnel to conduct the drill.

STANDARDS: Dismounted personnel immediately drop to prone positions and take the individual steps necessary to survive the blast. Mounted personnel immediately drop down inside the vehicle and brace themselves for the blast. After the blast wave, they establish security.

TASK STEPS AND PERFORMANCE MEASURES

1. All personnel react to an unwarranted nuclear attack. They take the following actions:

    a. **Dismounted** personnel immediately:

        (1) Drop to a prone position and close eyes.

        (2) Turn bodies head-on toward the blast.

        (3) Place thumbs in ears.

        (4) Cover faces with hands.

        (5) Place arms under bodies.

        (6) Tuck heads into shoulders; keeping helmets on and faces down.

        (7) Stay down until the blast wave passes and debris stops falling.

        (8) Check for casualties and damaged equipment.

    b. **Mounted** personnel immediately:

        (1) Drop down inside of vehicles.

        (2) Before the blast wave, close hatches.

        (3) Turn off unneeded radios.

        (4) Brace themselves.

2. After the blast wave has passed, personnel give first aid to and evacuate casualties.

3. The platoon/squad leader takes the following actions:

    a. Reestablishes the chain of command and communications.

    b. Submits an initial chemical, biological, radiological, and nuclear (CBRN) 1 report.

    c. Accounts for all Soldiers.

    d. Provides a situation report (SITREP) to higher HQ.

    e. Reorganizes as needed.

    f. Continues the mission.

SUPPORTING PRODUCTS

Product ID   **Product Name**
FM 3-21.8   The Infantry Rifle Platoon and Squad

SUPPORTING INDIVIDUAL TASKS

Task Number   **Task Title**
031-503-1018   React to Nuclear Hazard/Attack
031-503-1021   Mark NBC Contaminated Area

SUPPORTING COLLECTIVE TASKS

Task Number   **Task Title**
03-2-9224   Conduct Operational Decontamination

TASK: React to a Chemical Attack (03-3-D0035)

CONDITIONS: The element moves or is stationary, conducting operations. The unit is attacked with a chemical agent. Soldiers hear a chemical alarm, observe an unknown gas or liquid, or are ordered to don their protective mask.

CUE: Any Soldier gives an oral or visual signal for a chemical attack, or a chemical alarm activates.

STANDARDS: All soldiers don their protective mask within 9 seconds or 15 seconds for masks with a hood. Soldiers assume mission-oriented protection posture 4 (MOPP 4) within 8 minutes. The element identifies the chemical agent using M8 chemical detector paper and the M256 detector kit. The squad/platoon leader reports that the unit is under a chemical attack and submits chemical, biological, nuclear, radiological, and high-yield explosive (CBRNE) 1 reports to next higher echelon.

TASK STEPS AND PERFORMANCE MEASURES
1. Soldiers don their protective mask.
NOTE: If Soldiers are using Joint-Service, Lightweight, Integrated, Suit Technology (JSLIST), the hood is not on the mask. Soldiers are only allotted nine seconds to don their protective mask.
NOTE: The mask gives immediate protection against traditional warfare agents. The mask may not completely protect the Soldier from certain toxic industrial chemicals, but it provides the best available protection to enable him to evacuate the hazard area. He may be required to evacuate to a minimum safe distance of at least 300 meters upwind from the contamination (if possible) or as directed by the commander.
2. Soldiers give vocal or nonvocal alarm.
3. Within 60 seconds, Soldiers use the appropriate skin decontamination kit (SDK) for individual decontamination, as necessary.
4. Soldiers assume MOPP 4 within eight minutes.
5. Soldiers initiate self- or buddy-aid, as needed.
6. The element identifies the chemical agent using M8 chemical detector paper and the M256 detector kit.
7. The element leader reports the chemical attack to higher headquarters using the CBRNE 1 report.
8. Leaders determine if decontamination is required and requests support, if necessary.
9. The element initiates immediate decontamination within 15 minutes (if necessary).
10. If contamination is present, the squad/platoon marks the area before leaving.

TASK STEPS AND PERFORMANCE MEASURES
11. The element moves and displaces, as appropriate, or continues its mission.

SUPPORTING PRODUCTS
Product ID	Product Name
FM 3-11	Nuclear, Biological, and Chemical Defense Operation
FM 3-11.4	Nuclear, Biological, and Chemical (NBC) Protection

SUPPORTING INDIVIDUAL TASKS
Task Number	Task Title
031-503-1005	Submit and NBC 1 Report
031-503-1019	React to Chemical or Biological (CB) Hazard/Attack
031-503-1021	Mark NBC Contaminated Area
031-503-1031	Use the Chemical Agent Monitor
031-503-1042	Protect Yourself From Chemical and Biological (CB) Contamination Using Your Assigned Protective Mask
031-503-1037	Detect Chemical Agents Using M8 or M9 Detector Paper
071-326-0608	Use Visual Signaling Techniques
031-503-1042	Protect Yourself From CBRN Injury/Contamination When Changing MOPP (Using JSLIST)

SUPPORTING COLLECTIVE TASKS
Task Number	Task Title
03-2-9224	Conduct Operational Decontamination

Appendix A

Mortar Platoon Unit Task List

The unit task list (UTL) shown in Table A-1 identifies some of the collective tasks that the mortar platoon is organized, manned, and equipped to conduct according to their TOE. This list has been assembled to assist the platoon leader in developing his supporting collective task list to document which tasks to train to support the battalion METL. The platoon leader may accept risk and not train the entire UTL. The task numbers and task titles are listed under each of the six warfighting functions.

Table A-1. Mortar platoon unit task list

Task Number	Task Title
Mission Command	
07-2-5081	Conduct Troop-Leading Procedures (Platoon-Company)
55-2-4806	Prepare Equipment for Deployment
55-2-4828	Plan Unit Deployment Activities Upon Receipt of a WARNO
Movement & Maneuver	
07-3-1297	Defend a Mortar Unit Against a Ground Attack (Section-Platoon)
07-3-1306	Displace a Mortar Unit by Echelon (Section-Platoon)
07-3-1351	Occupy a Mortar Firing Position (Section-Platoon)
07-3-1360	Operate a Mortar Platoon by Split Sections and Squads (Section-Platoon)
07-3-2045	Reconnoiter a Mortar Firing Position (Section-Platoon)
07-2-3027	Integrate Direct Fires (Platoon-Company)
07-2-1396	Employ Obstacles (Platoon-Company)
19-3-2406	Conduct Roadblock and Checkpoint Operations
07-2-5027	Conduct Consolidation and Reorganization (Platoon-Company)
07-2-9014	Occupy an Assembly Area (Platoon-Company)

Table A-1. Mortar platoon unit task list (continued)

Task Number	Task Title
07-2-5009	Conduct a Rehearsal (Platoon-Company)
44-3-3220	Perform Passive Air Defense Measures
44-3-3221	Perform Active Air Defense Measures
07-2-1189	Conduct a Dismounted Tactical Road March (Platoon-Company)
07-2-6063	Maintain Operations Security (Platoon-Company)
07-2-1198	Conduct a Mounted Tactical Road March (Platoon-Company)
07-2-1342	Conduct Tactical Movement (Platoon-Company)
07-2-5036	Conduct Coordination (Platoon-Company)
07-2-6045	Employ Camouflage, Concealment, and Deception Techniques (Platoon-Company)
07-2-9005	Conduct a Linkup (Platoon-Company)
07-3-9013	Conduct Action on Contact
07-2-9011	Conduct Tactical Movement in an Urban Area (Platoon-Company)
07-2-1495	Conduct an Air Assault (Platoon-Company)
07-3-9017	Conduct Actions at Danger Areas
03-2-9226	Cross a Chemically Contaminated Area
Intelligence	
34-2-0010	Report Tactical Information
34-3-0001	Monitor Platoon Operational Status
34-3-0003	Maintain Operations Security
Fires	
07-2-3036	Integrate Indirect Fire Support (Platoon-Company)
07-3-3000	Adjust Mortar Final Protective Fire (Section-Platoon)
07-3-3018,	Fire Mortars by Direct Alignment (Section-Platoon)

Table A-1. Mortar platoon unit task list (continued)

Task Number	Task Title
07-3-3027	Fire a Mortar Final Protective Fire Mission (Section-Platoon)
07-3-3036	Adjust Mortar Fire during a Hipshoot Mission (Section-Platoon)
07-3-3045	Fire Immediate Suppression during a Hipshoot Mission (Section-Platoon)
07-3-3054	Fire a Mortar Adjust Fire Mission (Section-Platoon)
07-3-3063	Fire a Mortar Coordinated Illumination Mission (Section-Platoon)
07-3-3072	Fire a Mortar Fire for Effect Mission (Section-Platoon)
07-3-3081	Fire a Mortar Illumination Mission (Section-Platoon)
07-3-3090	Fire a Mortar Smoke Mission (Section-Platoon)
07-3-3099	Fire a Mortar Priority Target Mission (Section-Platoon)
07-3-3108	Fire a Mortar Quick Smoke Mission (Section-Platoon)
07-3-3117	Conduct Simultaneous Mortar Fire Missions (Section-Platoon)
07-3-3126	Fire a Time On Target Mortar Fire Mission (Section-Platoon)
07-3-3135	Register and Adjust a Parallel Mortar Sheaf (Section-Platoon)
07-3-5072	Operate a Mortar Fire Direction Center (Section-Platoon)
07-3-5090	Process a Mortar Call for Fire Mission (Section-Platoon)
Sustainment	
63-2-4546	Conduct Logistics Package (LOGPAC) Support
08-2-0003	Treat Casualties
08-2-0004	Evacuate Casualties
Protection	
03-2-9224	Conduct Operational Decontamination
07-2-4054	Secure Civilians During Operations (Platoon-Company)
03-2-9201	Implement CBRN Protective Measures
07-2-5063	Conduct Composite Risk Management (Platoon-Company)

This page intentionally left blank.

Appendix B

Battalion METL

Tables B-1, B-2, and B-3 are examples of the mortar platoon's battalion headquarters, (that of a CAB, within the HBCT [See Table B-1], and Infantry battalions within the Infantry [See Table B-2] and Stryker BCTs [See Table B-3]) METL.

Table B-1. Example of a CAB, HBCT METL

CAB, HBCT	MET (ART)
	TG (T&EO)
	Supporting Collective Task (T&EO)
ART 7.1.2	Conduct an Attack
07-6-1072	TG: Conduct a Movement to Contact (Battalion - Brigade)
07-6-1081	Conduct a Passage of Lines as Passing Unit (Battalion - Brigade)
07-6-6082	Conduct Mobility, Countermobility, and or Survivability (Battalion - Brigade)
07-6-1091	Conduct a Gap Crossing (Battalion - Brigade)
07-6-1252	Conduct a Combined Arms Breach of an Obstacle (Battalion - Brigade)
17-6-1007	Conduct Intelligence, Surveillance, and Reconnaissance (ISR) Synchronization and Integration (Battalion - Brigade)
71-8-2321	Develop the Intelligence, Surveillance, and Recon Plan
17-6-3004	Employ Fires and Effects (Battalion-Brigade)
07-6-5037	Conduct Consolidation (Battalion - Brigade)
07-6-5082	Conduct Reorganization (Battalion - Brigade)
63-1-4032	Coordinate LOGPAC Operations
71-8-5111	Conduct the Military Decision Making Process
71-8-5131	Execute Tactical Operations (Battalion - Corps)
71-8-5142	Evaluate Situation or Operation (Battalion - Corps)
03-2-9224	Conduct Operational Decontamination

Table B-1. Example of a CAB, HBCT METL (continued)

CAB, HBCT	MET (ART)
	TG (T&EO)
	Supporting Collective Task (T&EO)
07-6-1092	TG: Conduct an Attack (Battalion-Brigade)
07-6-1181	Conduct an Attack in an Urban Area (Battalion-Brigade)
07-6-1081	Conduct a Passage of Lines as Passing Unit (Battalion-Brigade)
07-6-6082	Conduct Mobility, Countermobility, and or Survivability (Battalion-Brigade)
07-6-1091	Conduct a Gap Crossing (Battalion-Brigade)
07-6-1252	Conduct a Combined Arms Breach of an Obstacle (Battalion-Brigade)
17-6-1007	Conduct Intelligence, Surveillance, and Reconnaissance (ISR) Synchronization and Integration (Battalion-Brigade)
71-8-2321	Develop the Intelligence, Surveillance, and Reconnaissance Plan
17-6-3004	Employ Fires and Effects (Battalion-Brigade)
07-6-5037	Conduct Consolidation (Battalion Brigade)
07-6-5082	Conduct Reorganization (Battalion-Brigade)
63-1-4032	Coordinate LOGPAC Operations
71-8-5111	Conduct the Military Decision Making Process
71-8-5131	Execute Tactical Operations (Battalion-Corps)
71-8-5142	Evaluate Situation or Operation (Battalion-Corps)
03-2-9224	Conduct Operational Decontamination
ART 7.2	**Conduct Defensive Operations**
07-6-1028	TG: Conduct a Defense (Battalion Brigade)
07-6-1036	Conduct a Delay (Battalion Brigade)
07-6-1144	Conduct a Withdrawal (Battalion Brigade)
07-6-1082	Conduct a Passage of Lines as Stationary Unit (Battalion Brigade)
07-6-6082	Conduct Mobility, Countermobility, and or Survivability (Battalion-Brigade)

Table B-1. Example of a CAB, HBCT METL (continued)

CAB, HBCT	MET (ART)
	TG (T&EO)
	Supporting Collective Task (T&EO)
17-6-1007	Conduct Intelligence, Surveillance, and Reconnaissance (ISR) Synchronization and Integration (Battalion-Brigade)
71-8-2321	Develop the Intelligence, Surveillance, and Reconnaissance Plan
17-6-3004	Employ Fires and Effects (Battalion-Brigade)
07-6-5037	Conduct Consolidation (Battalion-Brigade)
07-6-5082	Conduct Reorganization (Battalion-Brigade)
63-1-4032	Coordinate LOGPAC Operations
71-8-5111	Conduct the Military Decision Making Process
71-8-5131	Execute Tactical Operations (Battalion-Corps)
71-8-5142	Evaluate Situation or Operation (Battalion-Corps)
03-2-9224	Conduct Operational Decontamination
ART 6.7.3	**Conduct Security Operations**
17-6-9225	TG: Conduct a Screen (Battalion-Brigade)
07-6-1107	Conduct a Relief in Place (Battalion-Brigade)
17-6-1007	Conduct Intelligence, Surveillance, and Reconnaissance (ISR) Synchronization and Integration (Battalion-Brigade)
71-8-2321	Develop the Intelligence, Surveillance, and Recon Plan
17-6-3004	Employ Fires and Effects (Battalion-Brigade)
63-1-4032	Coordinate LOGPAC Operations
71-8-5111	Conduct the Military Decision Making Process
71-8-5131	Execute Tactical Operations (Battalion-Corps)
71-8-5142	Evaluate Situation or Operation (Battalion Corps)
17-6-9222	TG: Conduct a Guard (Battalion-Corps)
07-6-1082	Conduct a Passage of Lines as Stationary Unit (Battalion-Brigade)

Table B-1. Example of a CAB, HBCT METL (continued)

CAB, HBCT	MET (ART)
	TG (T&EO)
	Supporting Collective Task (T&EO)
17-6-3809	Conduct Battle Handover (Battalion-Brigade)
07-6-1107	Conduct a Relief in Place (Battalion-Brigade)
07-6-6082	Conduct Mobility, Countermobility, and or Survivability (Battalion-Brigade)
17-6-1007	Conduct Intelligence, Surveillance, and Reconnaissance (ISR) Synchronization and Integration (Battalion-Brigade)
71-8-2321	Develop the Intelligence, Surveillance, and Reconnaissance Plan
17-6-3004	Employ Fires and Effects (Battalion-Brigade)
63-1-4032	Coordinate LOGPAC Operations
71-8-5111	Conduct the Military Decision Making Process
71-8-5131	Execute Tactical Operations (Battalion-Corps)
71-8-5142	Evaluate Situation or Operation (Battalion-Corps)
07-6-1272	TG: Conduct Area Security (Battalion-Brigade)
17-6-9406	Conduct Lines of Communication Security (Battalion-Brigade)
19-1-2007	Conduct Convoy Security Operations
07-6-1082	Conduct a Passage of Lines as Stationary Unit (Battalion-Brigade)
07-6-1107	Conduct a Relief in Place (Battalion-Brigade)
07-6-6082	Conduct Mobility, Countermobility, and or Survivability (Battalion-Brigade)
17-6-1007	Conduct Intelligence, Surveillance, and Reconnaissance (ISR) Synchronization and Integration (Battalion-Brigade)
71-8-2321	Develop the Intelligence, Surveillance, and Reconnaissance Plan
17-6-3004	Employ Fires and Effects (Battalion-Brigade)
63-1-4032	Coordinate LOGPAC Operations

Table B-1. Example of a CAB, HBCT METL (continued)

CAB, HBCT	MET (ART)
	TG (T&EO)
	Supporting Collective Task (T&EO)
71-8-5111	Conduct the Military Decision Making Process
71-8-5131	Execute Tactical Operations (Battalion-Corps)
71-8-5334	Plan Public Affairs Operations (Battalion-Corps)
07-6-4000	Conduct a Civil Military Operation (Battalion - Brigade)
07-6-6073	Secure Civilians During Operations (Battalion - Brigade)
03-2-9224	Conduct Operational Decontamination
ART 7.3	**Conduct Stability Operations**
71-8-7331	TG: Coordinate Essential Services for Host Nation (Brigade - Corps)
17-6-9406	Conduct Lines of Communication Security (Battalion-Brigade)
19-1-2007	Conduct Convoy Security Operations
07-6-1272	Conduct Area Security (Battalion-Brigade)
17-6-1007	Conduct Intelligence, Surveillance, and Reconnaissance (ISR) Synchronization and Integration (Battalion-Brigade)
71-8-2321	Develop the Intelligence, Surveillance, and Reconnaissance Plan
17-6-3004	Employ Fires and Effects (Battalion-Brigade)
63-1-4032	Coordinate LOGPAC Operations
71-8-5111	Conduct the Military Decision Making Process
71-8-5131	Execute Tactical Operations (Battalion-Corps)
71-8-5142	Evaluate Situation or Operation (Battalion-Corps)
71-8-5334	Plan Public Affairs Operations (Battalion-Corps)
07-6-4000	Conduct a Civil Military Operation (Battalion-Brigade)
07-6-6073	Secure Civilians During Operations (Battalion-Brigade)

Table B-2. Example of Infantry battalion, IBCT METL

IN BN, IBCT	MET (ART)
	TG (T&EO)
	Supporting Collective Task (T&EO)
ART 7.1.2	**Conduct an Attack**
07-6-1072	TG: Conduct a Movement to Contact (Battalion-Brigade)
07-6-1081	Conduct a Passage of Lines as Passing Unit (Battalion-Brigade)
07-6-6082	Conduct Mobility, Countermobility, and or Survivability (Battalion-Brigade)
17-6-1007	Conduct Intelligence, Surveillance, and Reconnaissance (ISR) Synchronization and Integration (Battalion-Brigade)
71-8-2321	Develop the Intelligence, Surveillance, and Reconnaissance Plan
17-6-3004	Employ Fires and Effects (Battalion-Brigade)
07-6-5037	Conduct Consolidation (Battalion-Brigade)
07-6-5082	Conduct Reorganization (Battalion-Brigade)
63-1-4032	Coordinate LOGPAC Operations
71-8-5111	Conduct the Military Decision Making Process
71-8-5131	Execute Tactical Operations (Battalion-Corps)
71-8-5142	Evaluate Situation or Operation (Battalion-Corps)
03-2-9224	Conduct Operational Decontamination
07-6-1092	TG: Conduct an Attack (Battalion-Brigade)
07-6-1181	Conduct an Attack in an Urban Area (Battalion-Brigade)
07-6-1081	Conduct a Passage of Lines as Passing Unit (Battalion - Brigade)
07-6-6082	Conduct Mobility, Countermobility, and or Survivability (Battalion-Brigade)
17-6-1007	Conduct Intelligence, Surveillance, and Reconnaissance (ISR) Synchronization and Integration (Battalion-Brigade)
71-8-2321	Develop the Intelligence, Surveillance, and Reconnaissance Plan

Table B-2. Example of Infantry battalion, IBCT METL (continued)

IN BN, IBCT	MET (ART)
	TG (T&EO)
	Supporting Collective Task (T&EO)
17-6-3004	Employ Fires and Effects (Battalion-Brigade)
07-6-5037	Conduct Consolidation (Battalion-Brigade)
07-6-5082	Conduct Reorganization (Battalion-Brigade)
63-1-4032	Coordinate LOGPAC Operations
71-8-5111	Conduct the Military Decision Making Process
71-8-5131	Execute Tactical Operations (Battalion-Corps)
71-8-5142	Evaluate Situation or Operation (Battalion-Corps)
03-2-9224	Conduct Operational Decontamination
ART 7.2	***Conduct Defensive Operations***
07-6-1028	TG: Conduct a Defense (Battalion-Brigade)
07-6-1036	Conduct a Delay (Battalion-Brigade)
07-6-1144	Conduct a Withdrawal (Battalion-Brigade)
07-6-1082	Conduct a Passage of Lines as Stationary Unit (Battalion Brigade)
07-6-1107	Conduct a Relief in Place (Battalion-Brigade)
07-6-6082	Conduct Mobility, Countermobility, and or Survivability (Battalion-Brigade)
17-6-1007	Conduct Intelligence, Surveillance, and Reconnaissance (ISR) Synchronization and Integration (Battalion-Brigade)
71-8-2321	Develop the Intelligence, Surveillance, and Recon Plan
17-6-3004	Employ Fires and Effects (Battalion-Brigade)
07-6-5037	Conduct Consolidation (Battalion-Brigade)
07-6-5082	Conduct Reorganization (Battalion-Brigade)
63-1-4032	Coordinate LOGPAC Operations
71-8-5111	Conduct the Military Decision Making Process
71-8-5131	Execute Tactical Operations (Battalion-Corps)
71-8-5142	Evaluate Situation or Operation (Battalion-Corps)

Table B-2. Example of Infantry battalion, IBCT METL (continued)

IN BN, IBCT	MET (ART) TG (T&EO) Supporting Collective Task (T&EO)
03-2-9224	Conduct Operational Decontamination
ART 6.7.3	**Conduct Security Operations**
17-6-9225	TG: Conduct a Screen (Battalion-Brigade)
07-6-1082	Conduct a Passage of Lines as Stationary Unit (Battalion-Brigade)
17-6-3809	Conduct Battle Handover (Battalion-Brigade)
07-6-1107	Conduct a Relief in Place (Battalion Brigade)
17-6-1007	Conduct Intelligence, Surveillance, and Reconnaissance (ISR) Synchronization and Integration (Battalion-Brigade)
71-8-2321	Develop the Intelligence, Surveillance, and Reconnaissance Plan
17-6-3004	Employ Fires and Effects (Battalion-Brigade)
63-1-4032	Coordinate LOGPAC Operations
71-8-5111	Conduct the Military Decision Making Process
71-8-5131	Execute Tactical Operations (Battalion-Corps)
71-8-5142	Evaluate Situation or Operation (Battalion-Corps)
07-6-1272	TG: Conduct Area Security (Battalion-Brigade)
17-6-9406	Conduct Lines of Communication Security (Battalion - Brigade)
07-6-1082	Conduct a Passage of Lines as Stationary Unit (Battalion-Brigade)
07-6-1107	Conduct a Relief in Place (Battalion-Brigade)
07-6-6082	Conduct Mobility, Countermobility, and or Survivability (Battalion-Brigade)
17-6-1007	Conduct Intelligence, Surveillance, and Reconnaissance (ISR) Synchronization and Integration (Battalion-Brigade)
71-8-2321	Develop the Intelligence, Surveillance, and Reconnaissance Plan
17-6-3004	Employ Fires and Effects (Battalion-Brigade)
63-1-4032	Coordinate LOGPAC Operations

Table B-2. Example of Infantry battalion, IBCT METL (continued)

IN BN, IBCT	MET (ART)
	TG (T&EO)
	Supporting Collective Task (T&EO)
71-8-5111	Conduct the Military Decision Making Process
71-8-5131	Execute Tactical Operations (Battalion-Corps)
71-8-5142	Evaluate Situation or Operation (Battalion-Corps)
71-8-5334	Plan Public Affairs Operations (Battalion-Corps)
07-6-4000	Conduct a Civil Military Operation (Battalion-Brigade)
07-6-6073	Secure Civilians During Operations (Battalion-Brigade)
03-2-9224	Conduct Operational Decontamination
ART 7.3	**Conduct Stability Operations**
71-8-7331	TG: Coordinate Essential Services for Host Nation (Brigade-Corps)
17-6-9406	Conduct Lines of Communication Security (Battalion-Brigade)
07-6-1272	Conduct Area Security (Battalion-Brigade)
17-6-1007	Conduct Intelligence, Surveillance, and Reconnaissance (ISR) Synchronization and Integration (Battalion-Brigade)
71-8-2321	Develop the Intelligence, Surveillance, and Reconnaissance Plan
17-6-3004	Employ Fires and Effects (Battalion-Brigade)
63-1-4032	Coordinate LOGPAC Operations
71-8-5111	Conduct the Military Decision Making Process
71-8-5131	Execute Tactical Operations (Battalion-Corps)
71-8-5142	Evaluate Situation or Operation (Battalion-Corps)
71-8-5334	Plan Public Affairs Operations (Battalion-Corps)
07-6-4000	Conduct a Civil Military Operation (Battalion-Brigade)
07-6-6073	Secure Civilians During Operations (Battalion-Brigade)

Table B-3. Example of Infantry battalion, SBCT METL

IN BN, SBCT	MET (ART)
	TG (T&EO)
	Supporting Collective Task (T&EO)
ART 7.1.2	Conduct an Attack
07-6-1072	TG: Conduct a Movement to Contact (Battalion-Brigade)
07-6-1081	Conduct a Passage of Lines as Passing Unit (Battalion-Brigade)
07-6-6082	Conduct Mobility, Countermobility, and or Survivability (Battalion-Brigade)
07-6-1091	Conduct a Gap Crossing (Battalion - Brigade) (SBCT IN BN Only)
17-6-1007	Conduct Intelligence, Surveillance, and Reconnaissance (ISR) Synchronization and Integration (Battalion-Brigade)
71-8-2321	Develop the Intelligence, Surveillance, and Reconnaissance Plan
17-6-3004	Employ Fires and Effects (Battalion-Brigade)
07-6-5037	Conduct Consolidation (Battalion-Brigade)
07-6-5082	Conduct Reorganization (Battalion-Brigade)
63-1-4032	Coordinate LOGPAC Operations
71-8-5111	Conduct the Military Decision Making Process
71-8-5131	Execute Tactical Operations (Battalion-Corps)
71-8-5142	Evaluate Situation or Operation (Battalion-Corps)
03-2-9224	Conduct Operational Decontamination
07-6-1092	TG: Conduct an Attack (Battalion-Brigade)
07-6-1181	Conduct an Attack in an Urban Area (Battalion-Brigade)
07-6-1081	Conduct a Passage of Lines as Passing Unit (Battalion-Brigade)
07-6-6082	Conduct Mobility, Countermobility, and or Survivability (Battalion-Brigade)
07-6-1091	Conduct a Gap Crossing (Battalion-Brigade) (SBCT IN BN Only)

Table B-3. Example of Infantry battalion, SBCT METL (continued)

IN BN, SBCT	MET (ART)
	TG (T&EO)
	Supporting Collective Task (T&EO)
17-6-1007	Conduct Intelligence, Surveillance, and Reconnaissance (ISR) Synchronization and Integration (Battalion-Brigade)
71-8-2321	Develop the Intelligence, Surveillance, and Reconnaissance Plan
ART 7.1.2	**Conduct an Attack**
17-6-3004	Employ Fires and Effects (Battalion-Brigade)
07-6-5037	Conduct Consolidation (Battalion-Brigade)
07-6-5082	Conduct Reorganization (Battalion-Brigade)
63-1-4032	Coordinate LOGPAC Operations
71-8-5111	Conduct the Military Decision Making Process
71-8-5131	Execute Tactical Operations (Battalion-Corps)
71-8-5142	Evaluate Situation or Operation (Battalion-Corps)
03-2-9224	Conduct Operational Decontamination
ART 7.2	**Conduct Defensive Operations**
07-6-1028	TG: Conduct a Defense (Battalion-Brigade)
07-6-1036	Conduct a Delay (Battalion-Brigade)
07-6-1144	Conduct a Withdrawal (Battalion-Brigade)
07-6-1082	Conduct a Passage of Lines as Stationary Unit (Battalion-Brigade)
07-6-1107	Conduct a Relief in Place (Battalion-Brigade)
07-6-6082	Conduct Mobility, Countermobility, and or Survivability (Battalion-Brigade)
17-6-1007	Conduct Intelligence, Surveillance, and Reconnaissance (ISR) Synchronization and Integration (Battalion-Brigade)
71-8-2321	Develop the Intelligence, Surveillance, and Reconnaissance Plan
17-6-3004	Employ Fires and Effects (Battalion-Brigade)

Table B-3. Example of Infantry battalion, SBCT METL (continued)

IN BN, SBCT	MET (ART)
	TG (T&EO)
	Supporting Collective Task (T&EO)
07-6-5037	Conduct Consolidation (Battalion-Brigade)
07-6-5082	Conduct Reorganization (Battalion-Brigade)
63-1-4032	Coordinate LOGPAC Operations
71-8-5111	Conduct the Military Decision Making Process
71-8-5131	Execute Tactical Operations (Battalion-Corps)
ART 7.1.2	Conduct an Attack
71-8-5142	Evaluate Situation or Operation (Battalion-Corps)
03-2-9224	Conduct Operational Decontamination
ART 6.7.3	Conduct Security Operations
17-6-9225	TG: Conduct a Screen (Battalion-Brigade)
07-6-1082	Conduct a Passage of Lines as Stationary Unit (Battalion-Brigade)
17-6-3809	Conduct Battle Handover (Battalion-Brigade)
07-6-1107	Conduct a Relief in Place (Battalion-Brigade)
17-6-1007	Conduct Intelligence, Surveillance, and Reconnaissance (ISR) Synchronization and Integration (Battalion-Brigade)
71-8-2321	Develop the Intelligence, Surveillance, and Reconnaissance Plan
17-6-3004	Employ Fires and Effects (Battalion-Brigade)
63-1-4032	Coordinate LOGPAC Operations
71-8-5111	Conduct the Military Decision Making Process
71-8-5131	Execute Tactical Operations (Battalion-Corps)

Table B-3. Example of Infantry Battalion, SBCT METL (continued)

IN BN, SBCT	MET (ART)
	TG (T&EO)
	Supporting Collective Task (T&EO)
71-8-5142	Evaluate Situation or Operation (Battalion-Corps)
07-6-1272	TG: Conduct Area Security (Battalion-Brigade)
17-6-9406	Conduct Lines of Communication Security (Battalion-Brigade)
19-1-2007	Conduct Convoy Security Operations (SBCT IN BN Only)
07-6-1272	Conduct Area Security (Battalion-Brigade)
07-6-1082	Conduct a Passage of Lines as Stationary Unit (Battalion-Brigade)
07-6-1107	Conduct a Relief in Place (Battalion-Brigade)
07-6-6082	Conduct Mobility, Countermobility, and or Survivability (Battalion-Brigade)
17-6-1007	Conduct Intelligence, Surveillance, and Reconnaissance (ISR) Synchronization and Integration (Battalion-Brigade)
71-8-2321	Develop the Intelligence, Surveillance, and Reconnaissance Plan
17-6-3004	Employ Fires and Effects (Battalion-Brigade)
63-1-4032	Coordinate LOGPAC Operations
71-8-5111	Conduct the Military Decision Making Process
71-8-5131	Execute Tactical Operations (Battalion-Corps)
71-8-5142	Evaluate Situation or Operation (Battalion-Corps)
71-8-5334	Plan Public Affairs Operations (Battalion-Corps)
07-6-4000	Conduct a Civil Military Operation (Battalion-Brigade)
07-6-6073	Secure Civilians During Operations (Battalion-Brigade)
03-2-9224	Conduct Operational Decontamination
ART 7.3	Conduct Stability Operations
71-8-7331	TG: Coordinate Essential Services for Host Nation (Brigade-Corps)

Table B-3. Example of Infantry battalion, SBCT METL (continued)

IN BN, SBCT	MET (ART)
	TG (T&EO)
	Supporting Collective Task (T&EO)
17-6-9406	Conduct Lines of Communication Security (Battalion-Brigade)
19-1-2007	Conduct Convoy Security Operations
07-6-1272	Conduct Area Security (Battalion-Brigade)
17-6-1007	Conduct Intelligence, Surveillance, and Reconnaissance (ISR) Synchronization and Integration (Battalion-Brigade)
71-8-2321	Develop the Intelligence, Surveillance, and Reconnaissance Plan
17-6-3004	Employ Fires and Effects (Battalion-Brigade)
63-1-4032	Coordinate LOGPAC Operations
71-8-5111	Conduct the Military Decision Making Process
71-8-5131	Execute Tactical Operations (Battalion-Corps)
71-8-5142	Evaluate Situation or Operation (Battalion-Corps)
71-8-5334	Plan Public Affairs Operations (Battalion-Corps)
07-6-4000	Conduct a Civil Military Operation (Battalion-Brigade)
07-6-6073	Secure Civilians During Operations (Battalion-Brigade)

Appendix C

CATS Task Selection to Battalion METL Matrix

A CATS task selection to the battalion METL matrix is an example matrix containing the existing CATS task selections with which the mortar platoon may be required to train. These task selections enable the mortar platoon to conduct specific training that support METs and task groups of an Infantry battalion (See Table C-1) and a CAB (See Table C-2).

The contents of these example tables were assembled from existing CATS relating to the mortar platoon and are not complete. For more information regarding task selections relating to the mortar platoon, go to CATSs found in ATN and DTMS.

Table C-1. Example CATS task selection to battalion METL matrix

Infantry Battalion, BCT		METs and Task Groups					
		Attack		Defend	Security		Stability
Task Number	Task Title	Movement to Contact	Deliberate Attack	Area Defense	Screen	Area Security	Order & Safety
07-TS-1471	Conduct Battalion Operations	X	X	X	X	X	X
07-TS-1126	Conduct Offensive Operations	X	X				
07-TS-1127	Conduct Defensive Operations			X			
07-TS-1128	Conduct Security Operations				X	X	
07-TS-1129	Conduct Stability Operations						X
07-TS-1477	Execute Air Assault Operations	X	X				
07-TS-1054	Plan and Prepare for Operations	X	X	X	X	X	X
07-TS-1474	Sustain the Battalion	X	X	X	X	X	X
07-TS-1071	Deploy/Redeploy the Battalion	X	X	X	X	X	X
07-TS-1081	Perform Sustainment Section Operations	X	X	X	X	X	X
06-TS-1431	Plan, Coordinate and Supervise Fire Planning and Execution	X	X	X	X	X	X
06-TS-1432	Conduct Fire Planning	X	X	X	X	X	X

Table C-1. Example CATs task selection to battalion METL matrix (continued)

		METs and Task Groups					
		Attack		Defend	Security		Stability
Infantry Battalion, BCT							
Task Number	**Task Title**	Movement to Contact	Deliberate Attack	Area Defense	Screen	Area Security	Order & Safety
06-TS-1433	Prepare for Full Spectrum Operations	X	X	X	X	X	X
06-TS-1434	Conduct Targeting	X	X	X	X	X	X
07-TS-3477	Provide Mortar Support	X	X	X	X	X	X
07-TS-3478	Plan and Prepare for Operations (Mortars)	X	X	X	X	X	X
07-TS-3479	Protect the Unit (Mortars)	X	X	X	X	X	X
07-TS-3480	Move Tactically (Mortars)	X	X	X	X	X	X
07-TS-3486	Conduct Security Operations				X	X	
06-TS-4339	Conduct Fire Planning and Prepare for Operations	X	X	X	X	X	X
06-TS-4340	Conduct Occupation of the OP-FIST/COLT	X	X				
06-TS-4341	Execute Fire Missions	X	X	X	X	X	X
17-FS-4342	FIST/COLT Team Operations	X	X	X	X	X	X
07-TS-3490	Provide Health Services Support	X	X	X	X	X	X
71-TS-5500	Maintain Assigned Vehicle	X	X	X	X	X	X

Table C-2. Example CATs task selection to CAB METL matrix (continued)

Combined Arms Battalion, HBCT		METs and Task Groups							
		Attack		Defend		Security			Stability
Task Number	Task Title	Movement to Contact	Deliberate Attack	Area Defense	Screen	Guard	Area Security	Establish Public Order & Safety	
17-TS-3210	Employ Mortars	X	X	X	X	X	X	
71-TS-5100	Operate and Maintain Equipment	X	X	X	X	X	X	X
06-TS-4339	Conduct Fire Planning and Prepare for Operations	X	X	X	X	X	X	X
06-TS-4341	Execute Fire Missions	X	X	X	X	X	X	
06-TS-4342	Conduct FIST/COLT Team Operations	X	X	X	X	X	X	
71-TS-5015	Call for Fire	X	X	X	X	X	X	
06-TS-4340	Conduct Occupation of the OP-FIST/COLT			X	X	X		
17-TS-3106	Perform Basic Tactical Tasks-Platoon	X	X	X	X	X	X	
06-TS-4311	Plan Fire Support – Fires Cell	X	X	X	X	X	X	X
06-TS-4312	Prepare for Fire Support – Fires Cell	X	X	X	X	X	X	
06-TS-4313	Execute Fire Support – Fires Cell	X	X	X	X	X	X	
06-TS-4314	Analyze and Conduct Targeting	X	X	X	X	X	X	X
71-TS-1009	Plan and Prepare Operations	X	X	X	X	X	X	X

Table C-2. Example CATs task selection to CAB METL matrix (continued)

Combined Arms Battalion, HBCT		METs and Task Groups						
		Attack		Defend	Security			Stability
Task Number	Task Title	Movement to Contact	Deliberate Attack	Area Defense	Screen	Guard	Area Security	Establish Public Order & Safety
71-TS-1000	Conduct Combined Arms Battalion Operations	X	X	X	X	X	X	X
71-TS-1001	Conduct an Attack/Movement to Contact	X	X					
71-TS-1002	Conduct a Defense/Delay			X				
71-TS-1003	Conduct Security Operations (Screen, Guard, Area Security)				X	X	X	
71-TS-1004	Conduct Stability Operations							X
71-TS-1008	Conduct Deployment							
71-TS-1010	Conduct Sustainment Operations	X	X	X	X	X	X	X
71-TS-1077	Employ Fires	X	X	X	X	X	X	
71-TS-2120	Protect the Force	X	X	X	X	X	X	X

This page intentionally left blank.

Glossary

Acronym	Definition
1SG	first sergeant

A

AA	avenue of approach
AAR	after action report
AKO	Army Knowledge Online
AO	area of operation
ARFORGEN	Army forces generation
ATLDG	Army training and leader development guidance
ATN	Army Training Network
ATS	Army training strategy

B

BCIS	Battlefield Combat Identification System
BCT	brigade combat team
BCTC	battle command training center
BDA	battle damage assessment
BFSB	battlefield surveillance brigade
BHL	battle handover line
BN	battalion
BP	battle position
BRIDGEREP	bridge report

C

CAS	close air support
CASEVAC	casualty evacuation
CATS	combined arms training strategy
CB	chemical and biological
CBRN	chemical, biological, radiological, nuclear
CCIR	commanders critical information requirement
CCTT	close combat tactical trainer
CEF	contingency expeditionary force
COA	course of action
COMMO	communication
COMSEC	communications security
COP	common operational picture
CP	command post

Acronym	Definition
CTC	combat training center

D

DE	directed energy
DEF	deployment expeditionary force
DOTD	Directorate of Training and Doctrine
DTMS	Digital Training Management System

E

EA	engagement area
ECOA	enemy courses of action
EOD	explosive ordnance disposal
EOF	escalation of force
EPLRS	Enhanced Position Location Reporting System
EPW	enemy prisoners of war

F

FBCB2	Force XXI Battle Command Brigade and Below
FDC	fire direction center
FIST	fire support team
FM	field manual
FMC	fully mission capable
FO	forward observer
FPF	final protective fires
FPL	final protective lines
FRAGO	fragmentary order
FSC	forward support company
FTX	field training exercise

G

GSR	ground surveillance radar

H

HBCT	heavy brigade combat team
HPT	high payoff target
HQ	headquarters
HQDA	Headquarters, Department of the Army
HUMINT	human intelligence

Acronym	Definition
	I
IBCT	Infantry brigade combat team
IMINT	imagery intelligence
IP	internet protocol
IPB	intelligence preparation of the battlefield
ISR	intelligence, surveillance, and reconnaissance
	J
JCATS	Joint Conflict and Tactical Simulation System
	K
KIA	killed in action
	L
LD	line of departure
LDS	leader development strategy
LOA	limit of advance
LOGPAC	logistics package
LRP	logistics release point
LVCG	live, virtual, constructive, and gaming
	M
MCoE	Maneuver Center of Excellence
MEDEVAC	medical evacuation
MET	mission essential task
METL	mission-essential task list
METT-TC	mission, enemy, terrain and weather, troops and support available, time available and civil considerations
MILES	Multiple Integrated Laser Engagement System
MOPP 4	mission-oriented protective posture 4
MP	military police
MTF	medical treatment facility
MTOE	modified table of organization and equipment
MTP	mission training plan
MWD	military working dog

Acronym	Definition
	N
NBC	nuclear, biological, and chemical
NCO	noncommissioned officer
	O
OAKOC	observation, avenues of approach, key and decisive terrain, obstacles, and cover and concealment
OBSTINTEL	obstacle intelligence
OE	operational environment
OP	observation post
OPORD	operations order
	P
PERSTAT	personnel status
PDDE	power-driven decontamination equipment
PIO	police intelligence operations
PIR	priority intelligence requirements
PME	professional military education
PMESII-PT	political, military, economic, social, information, infrastructure, physical environment and time
POSNAV	position navigation
	R
RA	regular Army
R&S	reconnaissance and surveillance
RC	reserve component
REDCON	readiness condition
REMBASS	Remotely Monitored Battlefield Sensor System
ROI	rules of interaction
ROE	rules of engagement
RP	release point
	S
S-1	adjutant [Army]
S-4	supply officer [Army]
SBCT	Stryker brigade combat team
SCPE	simplified collective protective equipment
SIGINT	signal intelligence
SIR	specific information requirements

Acronym	Definition
SOEO	scheme of engineer operations
SOFA	status of forces agreement
SIR	specific information requirements
SITREP	situation report
SOI	signal operating instructions
SOP	standing operating procedure
SP	start point
SSI	signal supplemental instructions
STT	sergeants time training
STX	situation training exercise
SU	situational understanding

T

TADSS	training aids, devices, simulators, and simulations
T&EOS	training and evaluation outlines
TC	training circular
TDA	table of distribution and allowance
TLP	troop leading procedure
TOE	table of organization and equipment
TRADOC	Training and Doctrine Command
TRP	target reference point
TSOP	tactical standing operating procedure

U

UAS	Unmanned Aircraft System
UGS	unattended ground sensors
UTL	unit task list
UTM	unit training management

V

VBS2	Virtual Battlespace 2

W

WARNO	warning order
WFF	warfighting functions
WIA	wounded in action
WTPS	warfighter training support package

X

XO	executive officer

This page intentionally left blank.

References

SOURCES USED
These are the sources quoted or paraphrased in this publication.

ARMY PUBLICATIONS
ADP 3-0, *Unified Land Operations*, 10 October 2011.

ADP 4-0, *Sustainment*, 31 July 2012.

ADP 5-0, *The Operations Process*, 17 May 2012.

ADP 6-0, *Mission Command*, 17 May 2012.

ADP 7-0, *Training Units and Developing Leaders for Full Spectrum Operations*, 23 August 2012.

AR 190-8, *Enemy Prisoners of War, Retained Personnel, Civilian Internees and Other Detainees*, 1 October 1997

AR 350-1, *Army Training and Leader Development*, 18 December 2009.

AR 385-10, *The Army Safety Program*, 27 August 2007.

AR 600-8-1, *Army Casualty Program*, 30 April 2007.

ATP 4-11, *Army Motor Transport Operations*, 5 July 2013.

ATTP 3-18.12, *Air Assault Operations*, 1 March 2011.

ATTP 3-20.97, *Dismounted Reconnaissance Troop*, 16 November 2010.

ATTP 3-21.90, *Tactical Employment of Mortars*, 4 April 2011.

ATTP 3-21.9, *SBCT Infantry Rifle Platoon and Squad*, 8 December 2010.

ATTP 4-02, *Army Health System*, 7 October 2011.

FM 1-02, *Operational Terms and Graphics*, 21 September 2004.

FM 2-01.3, *Intelligence Preparation of the Battlefield/Battlespace*, 15 October 2009.

FM 3-07, *Stability Operations*, 6 October 2008.

FM 3-11, *Multiservice Doctrine for Chemical, Biological, Radiological, and Nuclear Operations*, 1 July 2011.

FM 3-11.4, *Multiservice Tactics, Techniques, and Procedures for Nuclear, Biological, and Chemical (NBC) Protection*, 2 June 2003.

FM 3-11.5, *Multiservice Tactics, Techniques, and Procedures for Chemical, Biological, Radiological, and Nuclear Decontamination*, 4 April 2006.

FM 3-20.15, *Tank Platoon*, 22 February 2007.

FM 3-20.96, *Reconnaissance and Cavalry Squadron*, 12 March 2010.

FM 3-20.971, *Reconnaissance and Cavalry Troop*, 4 August 2009.

FM 3-20.98, *Reconnaissance and Scout Platoon*, 3 August 2009.

FM 3-21.8, *The Infantry Rifle Platoon and Squad*, 28 March 2007.

FM 3-21.10, *The Infantry Rifle Company*, 27 July 2006.

FM 3-22.90, Mortars, 7 December 2007.

FM 3-28, *Civil Support Operations*, 20 August 2010.

FM 3-90.1, *Tank and Mechanized Infantry Company Team*, 9 December 2002.

FM 3-90.5, *The Combined Arms Battalion*, 7 April 2008.

FM 4-02.7, *Multiservice Tactics, Techniques, and Procedures for Health Service Support in a Chemical, Biological, Radiological, and Nuclear Environment*, 15 July 2009.

FM 4-25.11, *First Aid*, 23 December 2002.

FM 5-19, *Composite Risk Management*, 21 August 2006.

FM 6-22, *Army Leadership: Competent, Confident, and Agile*, 12 October 2006.

TC 3-34.489, *The Soldier and the Environment*, 8 May 2001.

ATLDG, Army, G-3/5/7 Memorandum, *Army Training and Leader Development Guidance, FY 10-11*, 31 July 2009. https://atn.army.mil/Media/docs/CSA%20ATLD%20GUIDANCE%202009.pdf

The Army Leader Development Strategy for a 21st Century Army, 25 November 2009. http://www.cgsc.edu/alds/ArmyLdrDevStrategy_20091125.pdf

JOINT AND DEPARTMENT OF DEFENSE PUBLICATIONS
JP 1-02, *Department of Defense (DOD) Dictionary of Military and Associated Terms*, 8 November 2010.

DOCUMENTS NEEDED

These documents must be available to the intended user of this publication.

DA Form 1156, *Casualty Feeder Card.*

DA Form 2028, *Recommended Changes to Publications and Blank Forms.*

DA Form 2188-R, *Data Sheet.*

DA Form 2188-1-R, *LHMBC/MFCS Data Sheet.*

DA Form 2399-R, *Computer's Record(LRA).*

DA Form 7566, *Composite Risk Management Worksheet.*

DD Form 1380, *U.S. Field Medical Card.*

READINGS RECOMMENDED
None

WEB SITES

Most Army doctrinal publications and regulations are available online at: http://www.apd.army.mil.

Army Knowledge Online (AKO) https://www.us.army.mil

Army Training Network (ATN) http://atn.army.mil/index.aspx

Combined Arms Training Strategy (CATS)
http://www.asat.army.mil/cats/catshome.htm

Digital Training Management System (DTMS) https://dtms.army.mil (individual password required)

MCoE Collective Training Branch Home Page
https://www.benning.army.mil

Most joint publications are available online at:
http://www.dtic.mil/doctrine/doctrine/doctrine.htm

This page intentionally left blank.

Index

This page intentionally left blank.

TC 3-21.90
2 AUGUST 2013

By Order of the Secretary of the Army:

RAYMOND T. ODIERNO
General, United States Army
Chief of Staff

Official:

GERALD B. O'KEEFE
Administrative Assistant to the
Secretary of the Army
1321401

DISTRIBUTION: *Active Army, Army National Guard, and U.S. Army Reserve*: To be distributed in accordance with the initial distribution number (IDN) 110801, requirements for TC 3-21.90

www.ingramcontent.com/pod-product-compliance
Lightning Source LLC
LaVergne TN
LVHW051504080426
835509LV00017B/1916